Welcome to *My Math*—your very own math book! You can write in it—in fact, you are encouraged to write, draw, circle, explain, and color as you explore the exciting world of mathematics. Let's get started. Grab a pencil and finish each sentence.

My name is ______________________.

My favorite color is ______________________.

My favorite hobby or sport is ______________________.

My favorite TV program or video game is

______________________.

My favorite class is ______________________.

mhmymath.com

STEM McGraw-Hill is committed to providing instructional materials in Science, Technology, Engineering, and Mathematics (STEM) that give all students a solid foundation, one that prepares them for college and careers in the 21st century.

Send all inquiries to:
McGraw-Hill Education
8787 Orion Place
Columbus, OH 43240

ISBN: 978-0-07-905760-0 (***Volume 2***)
MHID: 0-07-905760-8

Printed in the United States of America.

6 7 8 9 LWI 23 22 21 20

Grade 2 • Volume 2

Authors:

Carter • Cuevas • Day • Malloy
Altieri • Balka • Gonsalves • Grace • Krulik • Molix-Bailey
Moseley • Mowry • Myren • Price • Reynosa • Santa Cruz
Silbey • Vielhaber

GO digital ▶ connectED.mcgraw-hill.com

▶ Log In

1. Go to **connectED.mcgraw-hill.com.**

2. Log in using your username and password.

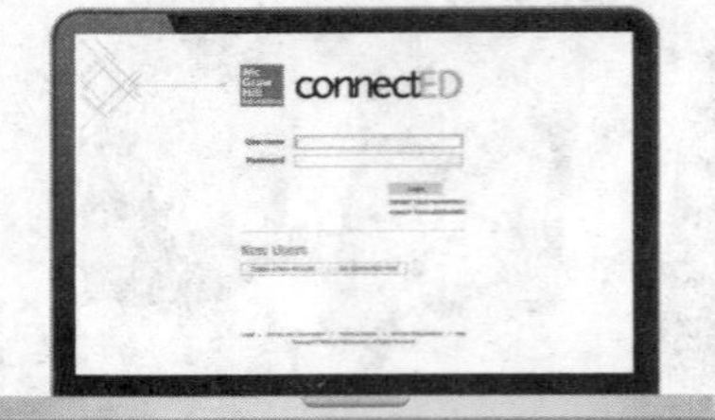

3. Click on the Student Edition icon to open the Student Center.

▶ Go to the Student Center

4. Click on Menu, then click on the **Resources** tab to see all of your online resources arranged by chapter and lesson.

5. Click on the **eToolkit** in the Lesson Resources section to open a library of eTools and virtual manipulatives.

6. Look here to find any assignments or messages from your teacher.

7. Click on the **eBook** to open your online Student Edition.

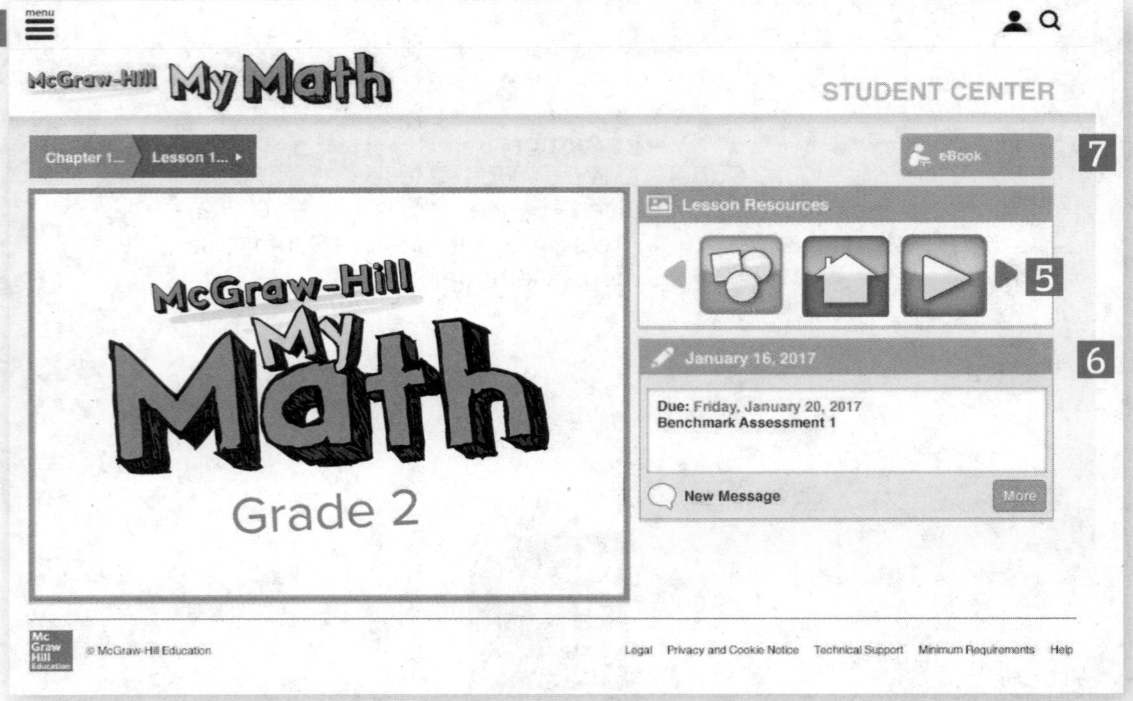

Username:

Password:

Explore the eBook!

8 Click the **speaker icon** at the top of the eBook pages to hear the page read aloud to you.

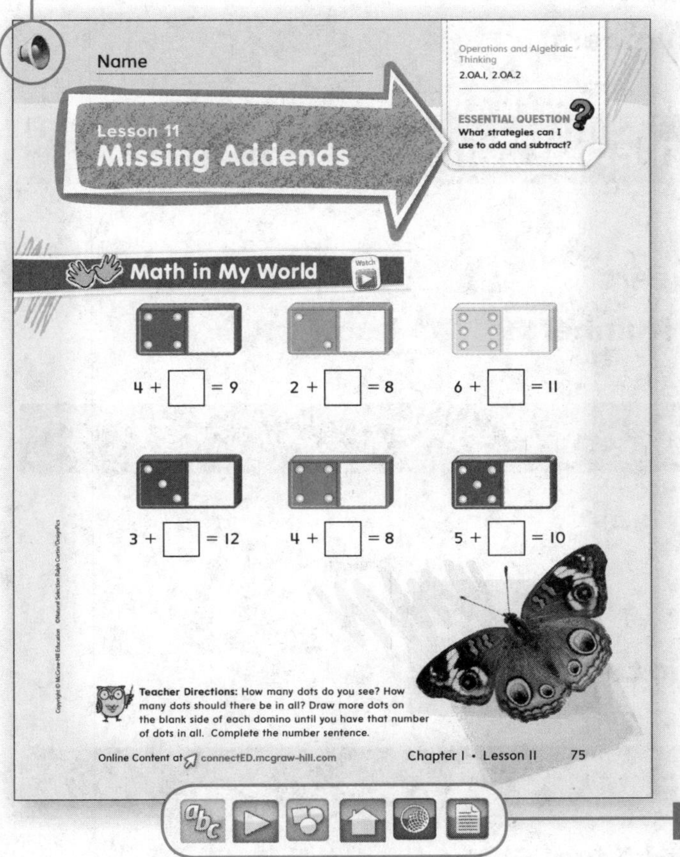
Name

Operations and Algebraic Thinking
2.OA.1, 2.OA.2

Lesson 11
Missing Addends

ESSENTIAL QUESTION
What strategies can I use to add and subtract?

Math in My World

4 + ☐ = 9
2 + ☐ = 8
6 + ☐ = 11
3 + ☐ = 12
4 + ☐ = 8
5 + ☐ = 10

Teacher Directions: How many dots do you see? How many dots should there be in all? Draw more dots on the blank side of each domino until you have that number of dots in all. Complete the number sentence.

Online Content at connectED.mcgraw-hill.com

Chapter 1 • Lesson 11 75

9 More resources can be found by clicking the icons at the bottom of the eBook pages.

 Practice and review your Vocabulary.

 Animations and videos allow you to explore mathematical topics.

 Explore concepts with eTools and virtual manipulatives.

 eHelp helps you complete your homework.

 Explore these fun digital activities to practice what you learned in the classroom.

 Worksheets are PDFs for Math at Home, Problem of the Day, and Fluency Practice.

Contents in Brief

Organized by Domain

Chapter

Apply Addition and Subtraction Concepts

Operations and Algebraic Thinking

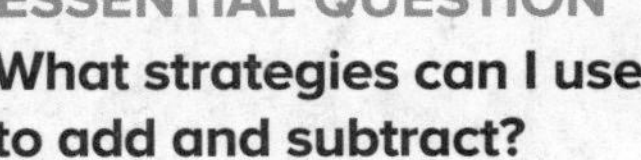

ESSENTIAL QUESTION

What strategies can I use to add and subtract?

Getting Started

Lessons and Homework

Wrap Up

There are Brain Builders problems in every lesson.

Chapter

Number Patterns

Operations and Algebraic Thinking

ESSENTIAL QUESTION
How can equal groups help me add?

Getting Started

Lessons and Homework

Wrap Up

Chapter 3

Add Two-Digit Numbers

Operations and Algebraic Thinking

ESSENTIAL QUESTION
How can I add two-digit numbers?

Getting Started

Lessons and Homework

Wrap Up

Chapter

4 Subtract Two-Digit Numbers

ESSENTIAL QUESTION

How can I subtract two-digit numbers?

Getting Started

Lessons and Homework

Wrap Up

Chapter 5 Place Value to 1,000

Number and Operations in Base Ten

ESSENTIAL QUESTION
How can I use place value?

Getting Started

Lessons and Homework

Wrap Up

Look for this!
Click online and you can find tools that will help you explore concepts.

Chapter

Add Three-Digit Numbers

Number and Operations in Base Ten

ESSENTIAL QUESTION

How can I add three-digit numbers?

Getting Started

Lessons and Homework

Wrap Up

Chapter 7

Subtract Three-Digit Numbers

Number and Operations in Base Ten

ESSENTIAL QUESTION

How can I subtract three-digit numbers?

Getting Started

Lessons and Homework

Wrap Up

Chapter

8 Money

Measurement and Data

ESSENTIAL QUESTION

How do I count and use money?

Getting Started

Lessons and Homework

Wrap Up

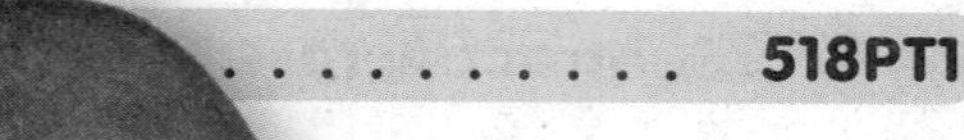

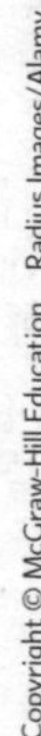

Chapter

Data Analysis

Measurement and Data

ESSENTIAL QUESTION

How can I record and analyze data?

Getting Started

Lessons and Homework

Wrap Up

Measurement and Data

ESSENTIAL QUESTION
How do I use and tell time?

Chapter 10 Time

Getting Started

Lessons and Homework

Wrap Up

Chapter

11 Customary and Metric Lengths

Measurement and Data

ESSENTIAL QUESTION

How can I measure objects?

Getting Started

Lessons and Homework

Wrap Up

Chapter 12

Geometric Shapes and Equal Shares

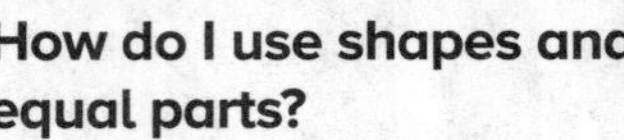

Geometry

ESSENTIAL QUESTION

How do I use shapes and equal parts?

Getting Started

Lessons and Homework

Wrap Up

Chapter

Subtract Three-Digit Numbers

ESSENTIAL QUESTION

How can I subtract three-digit numbers?

My School is Cool!

Watch a video!

Name

Chapter 7 Project

Reference Posters for Subtracting Three-Digit Numbers

1. Work together to determine how your group will show subtracting three-digit numbers on your poster.
2. Your plans should include a list of steps for strategies and drawings of concepts.
3. Plan an example of your poster in the space below.

My Vocabulary Cards

Vocab abc

Processes & Practices

Directions:
Ideas for Use

- Ask students to use the blank cards to draw or write words that will help them with concepts like three-digit subtraction or subtract across zeros.
- Have students use the blank cards to write basic subtraction facts. They should write the answer on the back of each card.

My Foldable

FOLDABLES® Follow the steps on the back to make your Foldable.

1

2

1 2 5

-1 0 9

3

1 2 5

-1 0 9

1 2 5

-1 0 9

Name ____________________

Independent Practice

Take apart hundreds to subtract.

3. 240 − 99

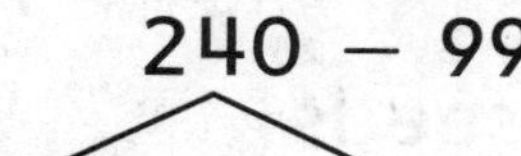

____ ____

____ − ____ = ____

____ + ____ = ____

So, 240 − 99 = ____.

4. 700 − 98

____ ____

____ − ____ = ____

____ + ____ = ____

So, 700 − 98 = ____.

5. 542 − 97

____ ____

____ − ____ = ____

____ + ____ = ____

So, 542 − 97 = ____.

6. 702 − 98

____ ____

____ − ____ = ____

____ + ____ = ____

So, 702 − 98 = ____.

7. 200 − 97

____ ____

____ − ____ = ____

____ + ____ = ____

So, 200 − 97 = ____.

8. 711 − 99

____ ____

____ − ____ = ____

____ + ____ = ____

So, 711 − 99 = ____.

Problem Solving

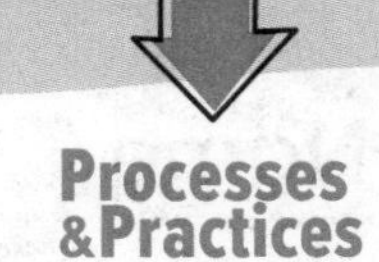

9. Eli counts 335 sunflower seeds. He places 98 of the seeds in a bowl. How many seeds are left?

_______ sunflower seeds

10. Rachel has 148 blueberries. She gives 97 blueberries to her brother. How many blueberries does Rachel have left?

_______ blueberries

Brain Builders

11. 382 people were in our school auditorium. 99 people left. How many people are still in the auditorium? Explain to a friend how you took apart hundreds to subtract in this problem.

_______ people

12. Explain in words how to solve 700 − 97.

____

____

____

Name ____________________

Lesson 1

Take Apart Hundreds to Subtract

My Homework

Homework Helper

Need help? connectED.mcgraw-hill.com

You can take apart hundreds to mentally subtract 324 − 99.

Take apart 324 as 224 and 100.

It is easier to subtract 99 from 100.

324 − 99

224 100

100 − 99 = 1

224 + 1 = 225

So, 324 − 99 = 225.

Practice

Take apart hundreds to subtract.

1. 835 − 98

_____ _____

_____ − _____ = _____

_____ + _____ = _____

So, 835 − 98 = _____.

2. 748 − 97

_____ _____

_____ − _____ = _____

_____ + _____ = _____

So, 748 − 97 = _____.

Take apart hundreds to subtract.

3. 395 − 99

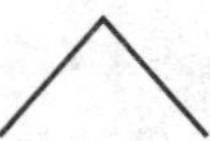

______ ______

______ − ______ = ______

______ + ______ = ______

So, 395 − 99 = ______.

4. 600 − 97

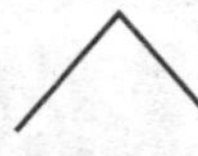

______ ______

______ − ______ = ______

______ + ______ = ______

So, 600 − 97 = ______.

Brain Builders

5. Students went to school for 180 days. 66 school days were sunny and 30 school days were cloudy. The rest of the school days were rainy. How many days were rainy?

______ rainy days

6. **Test Practice** How would you take apart 355 to solve 355 − 94?

○ 355 and 100 ○ 300 and 55

○ 255 and 100 ○ 255 and 94

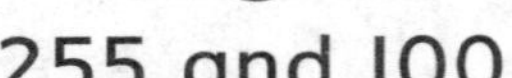

Math at Home Have your child solve 321 − 99 by taking apart hundreds to subtract.

Name

Lesson 2
Subtract Hundreds

ESSENTIAL QUESTION
How can I subtract three-digit numbers?

Math in My World

_______ steps

Teacher Directions: There are 400 steps in my school. My class decided to walk up all of the steps. We have walked up 300 steps already. How many steps do we have left? Use base-ten blocks. Draw the blocks. Write how many steps are left.

Guided Practice

You can use basic subtraction facts to help subtract hundreds. Find 500 − 300.

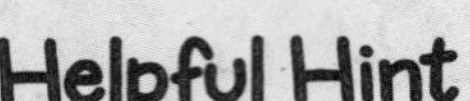

You know 5 − 3 = 2.
So, 500 − 300 = 200.

 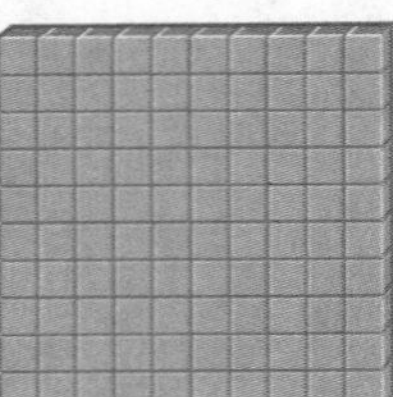

  5 hundreds
− 3 hundreds

 hundreds

500 − 300 = 200

Subtract.

1. 800 − 100 = ______

2. 200 − 200 = ______

3. 700 − 100

4. 400 − 200

5. 500 − 100

6. 700 − 300

Talk Math What subtraction fact can you use to find 900 − 800?

Name

Independent Practice

Helpful Hint
Use the subtraction facts you know to subtract hundreds.

Subtract.

7. 800 − 300 = ______

8. 600 − 600 = ______

9. 300 − 100 = ______

10. 900 − 200 = ______

11. 800 − 400 = ______

12. 700 − 500 = ______

13. 200 − 200 = ______

14. 400 − 100 = ______

15. 700 − 400 = ______

16. 800 − 200 = ______

17. 300 − 300 = ______

18. 500 − 400 = ______

19. 500 − 300 = ______

20. 200 − 100 = ______

21. 600 − 200 = ______

22. 900 − 700 = ______

23. 700 − 200 = ______

24. 400 − 300 = ______

25. 800 − 0 = ______

26. 800 − 800 = ______

Problem Solving

27. 700 people came to my school's spring concert. 500 people came to our winter concert. How many more people came to the spring concert?

_______ people

28. My school made flags for Memorial Day. There were 600 flags. 200 students took their flags home. How many flags are left?

_______ flags

Brain Builders

29. The floor in my classroom has 800 tiles. 100 tiles are blue and 300 are green. The rest of the tiles are white. How many white tiles are there? What facts did you use to solve this problem?

____

Write Math How is subtracting hundreds like subtracting ones?

____

____

____

____

Name ______________________

My Homework

Homework Helper

Need help? connectED.mcgraw-hill.com

Find 300 − 200.

Helpful Hint
You know 3 − 2 = 1.
So, 300 − 200 = 100.

  3 hundreds
− 2 hundreds
  1 hundred

300 − 200 = 100

Practice

Subtract.

1. 700 − 300 = ______

2. 500 − 200 = ______

3. $\begin{array}{r} 300 \\ -\ 100 \\ \hline \end{array}$

4. $\begin{array}{r} 100 \\ -\ 100 \\ \hline \end{array}$

5. $\begin{array}{r} 600 \\ -\ 400 \\ \hline \end{array}$

6. $\begin{array}{r} 800 \\ -\ 700 \\ \hline \end{array}$

7. $\begin{array}{r} 900 \\ -\ 100 \\ \hline \end{array}$

8. $\begin{array}{r} 500 \\ -\ 100 \\ \hline \end{array}$

9. $\begin{array}{r} 400 \\ -\ 400 \\ \hline \end{array}$

10. $\begin{array}{r} 400 \\ -\ 200 \\ \hline \end{array}$

Solve each word problem.

11. 800 people went to watch the ballet. 200 people left early. How many people stayed at the ballet?

_______ people

Brain Builders

12. 500 boys and girls went to the basketball game. 300 girls were at the game. How many boys were at the game? Write the number sentence you used to solve the problem.

____

13. The cafeteria had pizza, hamburgers, and hot dogs for lunch. 900 lunches were sold. 300 people bought hot dogs and 200 people bought hamburgers. How many people bought pizza?

_______ people

14. **Test Practice** Which number sentence could help you solve 700 − 500?

7 − 2	2 + 5	5 + 2	7 − 5
○	○	○	○

Math at Home Ask your child what number is 100 less than 500.

Name

Lesson 3

Mentally Subtract 10 or 100

ESSENTIAL QUESTION

How can I subtract three-digit numbers?

Math in My World

_______ students

Teacher Directions: 135 students are on the playground. 100 students go inside. How many students are left on the playground? Use base-ten blocks to solve. Write the number.

Guided Practice

To subtract 100 or 10, think of facts you know.
Mentally subtract 100.

$$\begin{array}{r} 422 \\ -\ 100 \\ \hline \end{array}$$

$$\begin{array}{r} 422 \\ -\ 100 \\ \hline 322 \end{array}$$

Mentally subtract 10.

$$\begin{array}{r} 460 \\ -\ 10 \\ \hline \end{array}$$

$$\begin{array}{r} 460 \\ -\ 10 \\ \hline 450 \end{array}$$

Subtract.

1. $\begin{array}{r} 329 \\ -\ 100 \\ \hline \end{array}$

2. $\begin{array}{r} 820 \\ -\ 10 \\ \hline \end{array}$

3. $\begin{array}{r} 363 \\ -\ 100 \\ \hline \end{array}$

4. $\begin{array}{r} 678 \\ -\ 10 \\ \hline \end{array}$

5. $\begin{array}{r} 724 \\ -\ 10 \\ \hline \end{array}$

6. $\begin{array}{r} 164 \\ -\ 10 \\ \hline \end{array}$

7. 900 − 100 = ________

8. 743 − 10 = ________

Talk Math Tell how to mentally subtract 10 or 100.

Name ....

Independent Practice

Subtract.

9. $\begin{array}{r} 491 \\ -\ 100 \\ \hline \end{array}$

10. $\begin{array}{r} 942 \\ -\ 10 \\ \hline \end{array}$

11. $\begin{array}{r} 770 \\ -\ 10 \\ \hline \end{array}$

12. $\begin{array}{r} 672 \\ -\ 100 \\ \hline \end{array}$

13. $\begin{array}{r} 853 \\ -\ 100 \\ \hline \end{array}$

14. $\begin{array}{r} 269 \\ -\ 10 \\ \hline \end{array}$

15. $\begin{array}{r} 368 \\ -\ 10 \\ \hline \end{array}$

16. $\begin{array}{r} 374 \\ -\ 100 \\ \hline \end{array}$

17. $\begin{array}{r} 982 \\ -\ 100 \\ \hline \end{array}$

18. 498 − 100 = ________

19. 533 − 100 = ________

Find the missing number.

20. 434 − ________ = 424

21. 371 − ________ = 271

22. 738 − ________ = 638

23. 270 − ________ = 260

Problem Solving

Processes & Practices

24. 863 flowers bloomed on the school playground. 100 of the flowers were eaten by bugs. How many flowers are on the playground now?

_________ flowers

25. 299 people came to the school play. 10 people had to leave early. How many people were still at the play?

_________ people

Brain Builders

26. 423 wildflowers grew in the field. Embry picked 10 wildflowers, and Faith picked 100. How many wildflowers are left in the field?

_________ wildflowers

Write Math Explain how you would mentally subtract 10 from 900.

____

____

____

Name

My Homework

Lesson 3

Mentally Subtract 10 or 100

Homework Helper

Need help? connectED.mcgraw-hill.com

Mentally subtract 100.

```
  567
- 100
  467
```

Helpful Hint
You know that
$5 - 1 = 4$.

Mentally subtract 10.

```
  567
-  10
  557
```

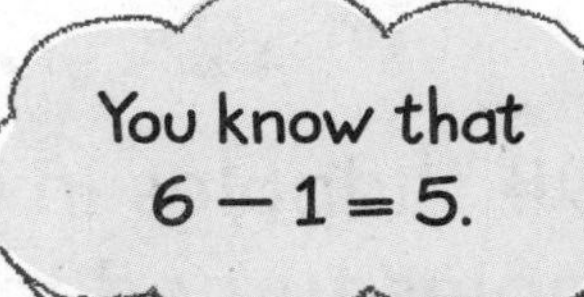

Practice

Subtract.

1. $477 - 100$

2. $378 - 10$

3. $879 - 100$

4. $245 - 10$

5. $849 - 100$

6. $320 - 10$

Find the missing numbers.

7. 358 − ______ = 258

8. 843 − ______ = 833

9. 954 − ______ = 944

10. 700 − ______ = 600

Brain Builders

11. Joan counts 543 birds by the pond. Some birds fly away. Now Joan counts 243 birds by the pond. How many birds flew away?

______ birds

12. There were 694 tadpoles in the lake. 110 of the tadpoles turned into frogs. How many tadpoles were left in the lake?

______ tadpoles

13. **Test Practice** Find 363 − 100.

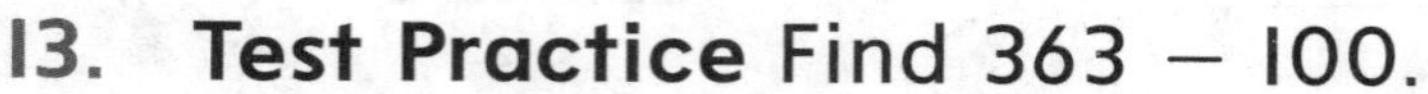

○ 163 ○ 263 ○ 363 ○ 463

Math at Home Practice saying a number and having your child subtract 10 or 100 from the number mentally.

Name ____________________

Vocabulary Check

Draw lines to match.

1. **difference**	To find the difference between two sets.
2. **subtract**	To take apart a number to write it in a new way.
3. **regroup**	The answer in a subtraction problem.

Concept Check

Take apart the hundreds to subtract.

4. 345 − 98

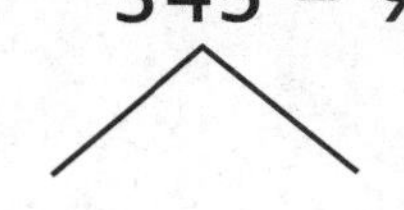

_____ _____

_____ − _____ = _____

_____ + _____ = _____

So, 345 − 98 = _____.

5. 926 − 99

_____ _____

_____ − _____ = _____

_____ + _____ = _____

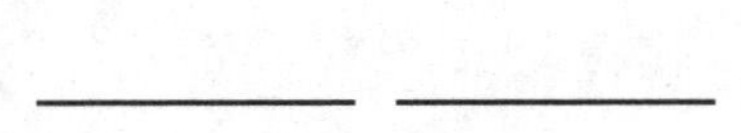

So, 926 − 99 = _____.

Subtract.

6. 900 − 400 = ______

7. 700 − 300 = ______

8. 393 − 100

9. 264 − 10

10. 737 − 100

Find the missing number.

11. 394 − ______ = 294

12. 842 − ______ = 832

13. 535 − ______ = 435

14. 253 − ______ = 243

Brain Builders

15. Brooklyn read 135 books over the summer.
Aubrey read 10 less books than Brooklyn.
Lila read 99 less books than Aubrey.
How many books did Lila read?

______ books

16. **Test Practice** Find 300 − 300.

600 ○ 300 ○ 100 ○ 0 ○

Name

Lesson 4
Regroup Tens

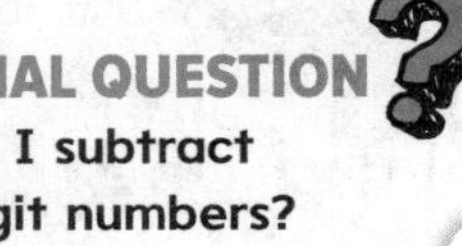

ESSENTIAL QUESTION
How can I subtract three-digit numbers?

Math in My World

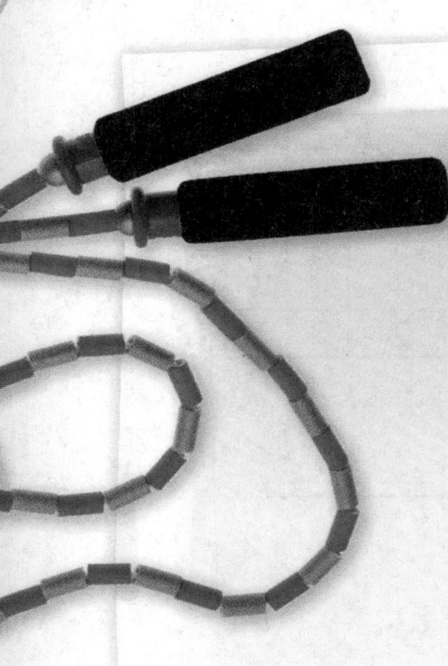

_______ jump ropes

Teacher Directions: Use base-ten blocks to solve. Mr. Hicks, our gym teacher, had 145 jump ropes. 36 jump ropes broke this year. How many jump ropes does Mr. Hicks have left? Write the number.

Guided Practice

Find 652 − 429.

Step 1 Subtract the ones. You cannot subtract 9 from 2. Regroup 1 ten as 10 ones.

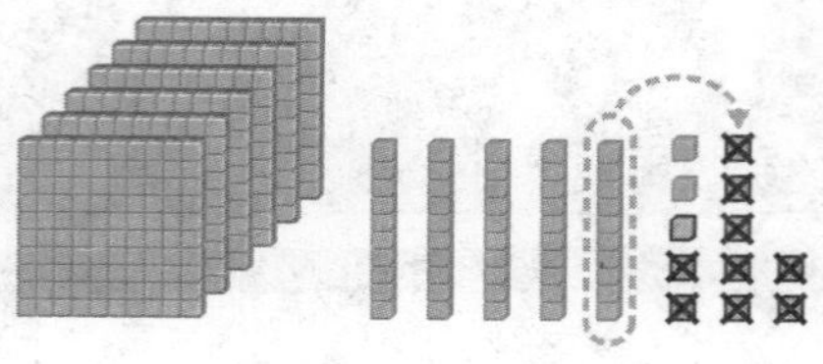

	hundreds	tens	ones
		4	12
	6	~~5~~	~~2~~
−	4	2	9
			3

Step 2 Subtract the tens.

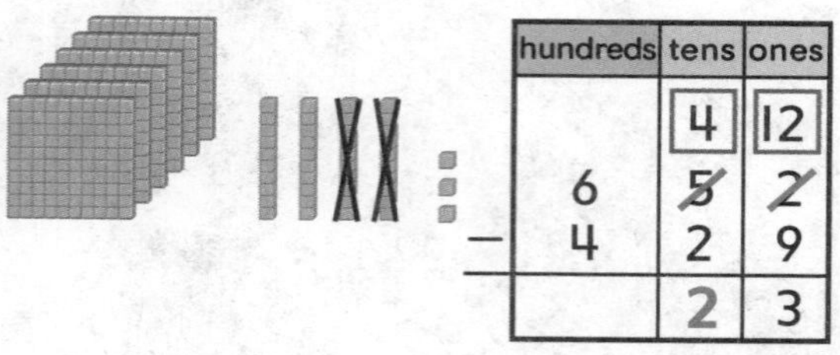

	hundreds	tens	ones
		4	12
	6	~~5~~	~~2~~
−	4	2	9
		2	3

Step 3 Subtract the hundreds.

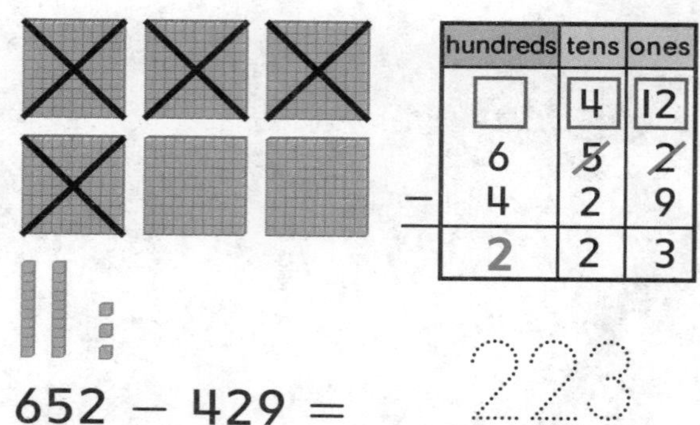

	hundreds	tens	ones
		4	12
	6	~~5~~	~~2~~
−	4	2	9
	2	2	3

652 − 429 = 223

Use Work Mat 7 and base-ten blocks. Subtract.

1.

	hundreds	tens	ones
	5	4	6
−		1	7

2.

	hundreds	tens	ones
	7	8	3
−	4	3	9

Talk Math How is subtracting three-digit numbers like subtracting two-digit numbers?

Name

Independent Practice

Use Work Mat 7 and base-ten blocks. Subtract.

3.

hundreds	tens	ones
	☐	☐
3	8	2
− 1	2	8

4.

hundreds	tens	ones
	☐	☐
4	6	7
−	4	9

5.

$$\begin{array}{r|r|r} & \square & \square \\ 5 & 7 & 5 \\ - & 6 & 6 \\ \hline & & \end{array}$$

6.

$$\begin{array}{r|r|r} & \square & \square \\ 8 & 6 & 3 \\ -2 & 1 & 8 \\ \hline & & \end{array}$$

7.

$$\begin{array}{r|r|r} & \square & \square \\ 2 & 6 & 4 \\ -1 & 3 & 5 \\ \hline & & \end{array}$$

8. $\begin{array}{r} 754 \\ -\ 507 \\ \hline \end{array}$

9. $\begin{array}{r} 455 \\ -\ 326 \\ \hline \end{array}$

10. $\begin{array}{r} 930 \\ -\ 428 \\ \hline \end{array}$

11. $\begin{array}{r} 780 \\ -\ 436 \\ \hline \end{array}$

12. $\begin{array}{r} 652 \\ -\ 35 \\ \hline \end{array}$

13. $\begin{array}{r} 931 \\ -\ 6 \\ \hline \end{array}$

14. $\begin{array}{r} 387 \\ -\ 18 \\ \hline \end{array}$

15. $\begin{array}{r} 423 \\ -\ 119 \\ \hline \end{array}$

16. $\begin{array}{r} 540 \\ -\ 15 \\ \hline \end{array}$

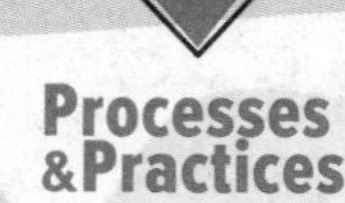

Problem Solving

17. Ada jumped 382 times in a row. Helen jumped 277 times in a row. How many more jumps in a row did Ada do than Helen?

_______ jumps

18. A store had 472 bouncy balls. They sold 155. How many bouncy balls were left?

_______ bouncy balls

Brain Builders

19. Lisa hit 294 golf balls in one day. She hit 149 in the morning. How many golf balls did she hit in the afternoon? Explain how you regrouped to a friend.

_______ golf balls

20. Andre wrote $381 - 165 = 224$.
Tell why Andre is wrong. Make it right.

____

____

____

Name ....

My Homework

Lesson 4

Regroup Tens

Homework Helper

eHelp

Need help? connectED.mcgraw-hill.com

Find 362 − 145.

Step 1 Subtract the ones. You cannot subtract 5 from 2. Regroup if needed.

Step 2 Subtract the tens.

Step 3 Subtract the hundreds.

	hundreds	tens	ones
		5	12
	3	~~6~~	~~2~~
−	1	4	5
	2	1	7

Practice

Subtract.

1.

	hundreds	tens	ones
	8	3	5
−	2	1	6

2.

	hundreds	tens	ones
	9	5	2
−	6	3	7

3.

	1	5	3
−		4	4

4.

	6	4	2
−	2	1	8

5.

	7	3	4
−	2	2	8

Subtract.

6. $\begin{array}{r} 153 \\ -\ \ 47 \\ \hline \end{array}$

7. $\begin{array}{r} 642 \\ -\ 215 \\ \hline \end{array}$

8. $\begin{array}{r} 754 \\ -\ 225 \\ \hline \end{array}$

9. $\begin{array}{r} 438 \\ -\ 129 \\ \hline \end{array}$

10. $\begin{array}{r} 362 \\ -\ 148 \\ \hline \end{array}$

11. $\begin{array}{r} 762 \\ -\ 349 \\ \hline \end{array}$

12. $\begin{array}{r} 647 \\ -\ 518 \\ \hline \end{array}$

13. $\begin{array}{r} 377 \\ -\ 163 \\ \hline \end{array}$

Brain Builders

14. 755 people came to my school's pancake breakfast. 318 of those people were children. 28 were teachers. The rest were parents. How many parents were at the breakfast?

_______ parents

15. **Test Practice** Which subtraction problem needs regrouping to solve?

392 − 222 ○

385 − 266 

692 − 321 ○

295 − 172 

Math at Home Write a subtraction problem that requires regrouping tens on a piece of paper for your child to solve. Have your child explain each step to solving the problem. Try one-, two-, and three-digit subtraction problems.

Name

Lesson 5
Regroup Hundreds

ESSENTIAL QUESTION
How can I subtract three-digit numbers?

 Math in My World  

_____ pieces of macaroni

Teacher Directions: Use base-ten blocks to solve. Gwen put 214 pieces of macaroni on her picture in art class. 120 pieces fell off. How many pieces of macaroni are still on the picture? Draw the blocks you used. Write the number.

Processes
&Practices

Guided Practice

Find 539 − 285.

Step 1
Subtract the ones.

hundreds	tens	ones
5	3	9
− 2	8	5
		4

Step 2
Subtract the tens.
You cannot subtract 8 from 3. Regroup 1 hundred as 10 tens.

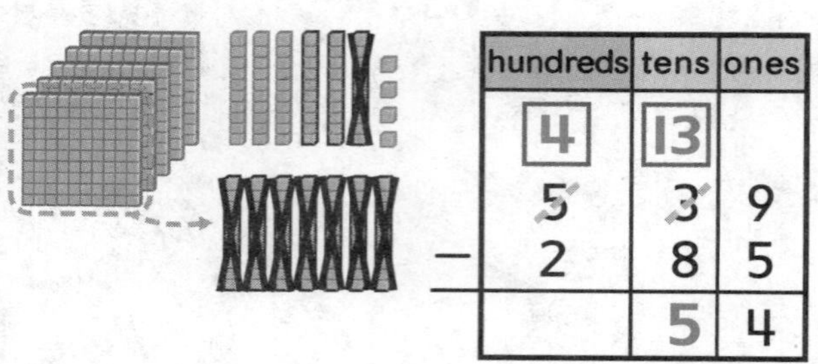

hundreds	tens	ones
4	13	
~~5~~	~~3~~	9
− 2	8	5
	5	4

Step 3
Subtract the hundreds.

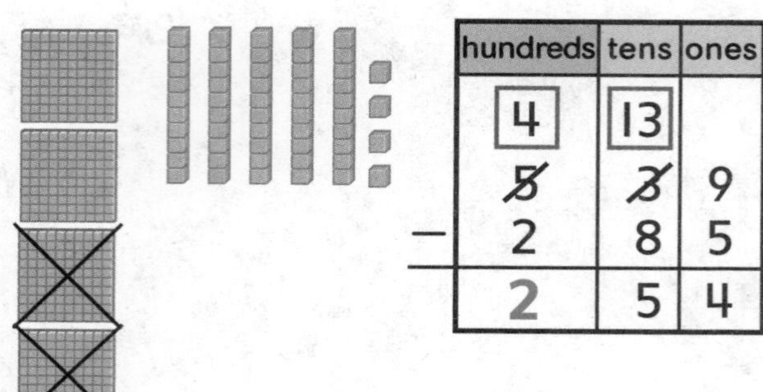

hundreds	tens	ones
4	13	
~~5~~	~~3~~	9
− 2	8	5
2	5	4

So, 539 − 285 = 254

Use Work Mat 7 and base-ten blocks. Subtract.

1.

hundreds	tens	ones
☐	☐	
5	6	8
− 1	9	7

2.

hundreds	tens	ones
☐	☐	
6	2	8
− 4	4	2

Talk Math How do you know when to regroup?

Name ______________________

Independent Practice

Use Work Mat 7 and base-ten blocks. Subtract.

3.

	hundreds	tens	ones
	□	□	
	5	8	6
−	2	9	5

4.

	hundreds	tens	ones
	□	□	
	6	3	8
−		4	3

5.

	□	□	
	4	5	9
−		6	9

6.

	□	□	
	7	3	9
−	5	4	1

7.

	□	□	
	8	2	7
−	2	4	7

8. $638 - 36$

9. $232 - 170$

10. $948 - 472$

11. $565 - 272$

12. $640 - 50$

13. $729 - 135$

14. $225 - 133$

15. $485 - 194$

16. $529 - 395$

Problem Solving

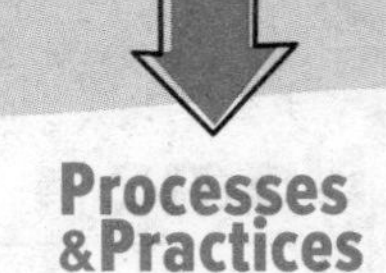

17. Lucy went to basketball camp. 325 second graders and 234 first graders were there. How many more second graders were at camp?

_______ more second graders

18. 417 boys were at the museum. 286 girls were at the museum. How many more boys were at the museum than girls?

_______ more boys

Brain Builders

19. 475 parents came to the school's fall party. 264 had to leave early. Would you need to regroup hundreds to find how many parents are still at the fall party? Explain.

____

____

____

Write Math Explain when and how to regroup hundreds.

____

____

____

____

Name

Lesson 6
Subtract Three-Digit Numbers

ESSENTIAL QUESTION
How can I subtract three-digit numbers?

Math in My World

Moo juice!

______ students

Teacher Directions: Use base-ten blocks to solve. 355 students are eating lunch in the cafeteria. 166 students packed their lunch. How many students bought their lunch? Draw the blocks. Write the number.

Guided Practice

Find 634 − 159.

Step 1 Subtract the ones. You cannot subtract 9 ones from 4 ones. Regroup 1 ten as 10 ones.

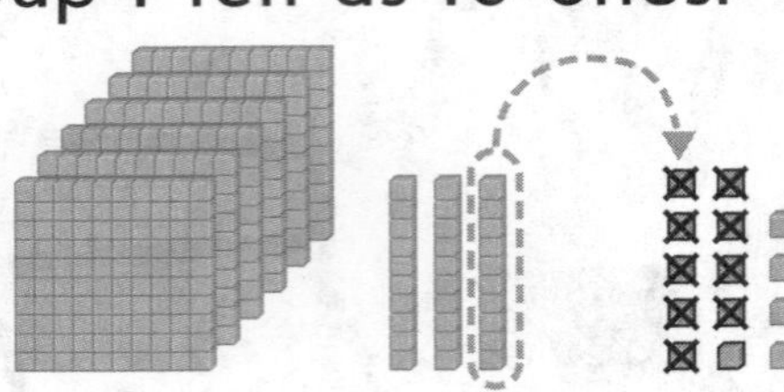

	hundreds	tens	ones
	☐	2	14
	6	~~3~~	~~4~~
−	1	5	9
			5

Step 2 Subtract the tens. You cannot subtract 5 tens from 2 tens. Regroup 1 hundred as 10 tens.

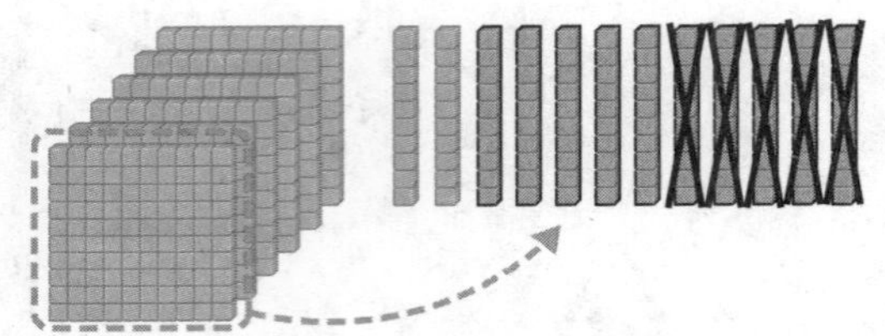

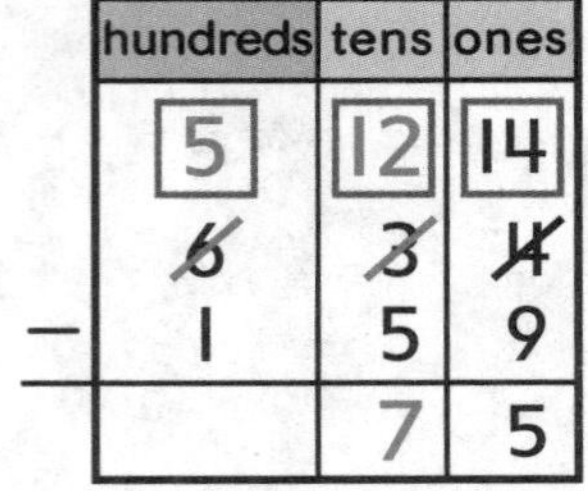

	hundreds	tens	ones
	5	12	14
	~~6~~	~~3~~	~~4~~
−	1	5	9
		7	5

Step 3 Subtract the hundreds. Now you can subtract 1 hundred from 5 hundreds.

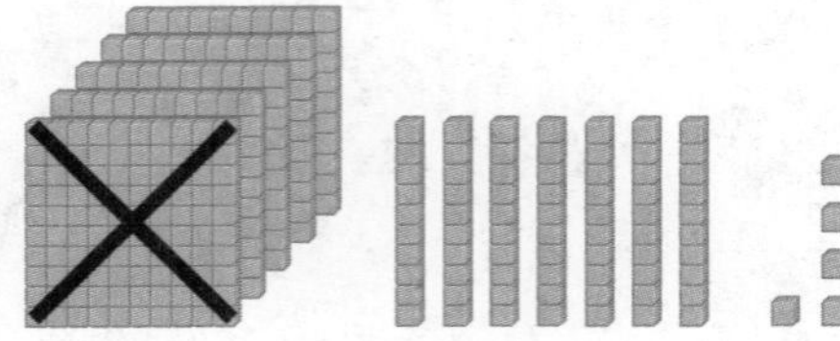

	hundreds	tens	ones
	5	12	14
	~~6~~	~~3~~	~~4~~
−	1	5	9
	4	7	5

634 − 159 = ______

Talk Math Explain what you write in the box above the ones and the box above the tens when you regroup tens and hundreds.

Name ________________________

Independent Practice

Use Work Mat 7 and base-ten blocks. Subtract.

1.

☐	☐	☐
4	3	1
− 3	4	5

2.

☐	☐	☐
7	6	2
−	8	8

3.

☐	☐	☐
9	5	3
− 7	6	4

4. $\begin{array}{r} 702 \\ -\ 211 \\ \hline \end{array}$

5. $\begin{array}{r} 884 \\ -\ 197 \\ \hline \end{array}$

6. $\begin{array}{r} 632 \\ -\ 444 \\ \hline \end{array}$

7. $\begin{array}{r} 485 \\ -\ 296 \\ \hline \end{array}$

8. $\begin{array}{r} 357 \\ -\ 169 \\ \hline \end{array}$

9. $\begin{array}{r} 625 \\ -\ 438 \\ \hline \end{array}$

10. $\begin{array}{r} 590 \\ -\ 184 \\ \hline \end{array}$

11. $\begin{array}{r} 718 \\ -\ 628 \\ \hline \end{array}$

12. $\begin{array}{r} 394 \\ -\ 185 \\ \hline \end{array}$

13. $\begin{array}{r} 561 \\ -\ 273 \\ \hline \end{array}$

14. $\begin{array}{r} 934 \\ -\ 395 \\ \hline \end{array}$

15. $\begin{array}{r} 533 \\ -\ 203 \\ \hline \end{array}$

Problem Solving

Processes & Practices

16. The post office had 912 stamps on Friday. By Saturday 189 stamps were left. How many stamps were sold on Friday?

_______ stamps

17. Dakota had 200 plastic bugs. He lost some bugs. He has 155 bugs left. How many bugs did he lose?

_______ bugs

Brain Builders

18. There are 550 pieces of popcorn. Ayana ate 177 pieces. Gabby ate 198 pieces. How many pieces did Ayana and Gabby eat all together? How many pieces are left?

____

19. 228 students like red. 293 students like blue. 154 students like green. How many more students like red or green than blue? Explain.

____

____

Name

My Homework

Lesson 6

Subtract Three-Digit Numbers

Homework Helper

Need help? connectED.mcgraw-hill.com

Sometimes when you subtract, you need to regroup the tens and hundreds.

Find 621 − 475.

Step 1 Subtract the ones. You cannot subtract 5 from 1. Regroup 1 ten as 10 ones.

Step 2 Subtract the tens. You cannot subtract 7 from 1. Regroup.

Step 3 Subtract the hundreds.

	hundreds	tens	ones
	5	11	11
	~~6~~	~~2~~	~~1~~
−	4	7	5
	1	4	6

Practice

Subtract.

1.

	4	5	3
−	3	5	1

2.

	6	1	2
−	1	5	9

Subtract.

3.

☐	☐	☐
5	3	7
−	2	6

4.

☐	☐	☐
6	3	4
− 2	7	8

5.

☐	☐	☐
3	6	4
− 1	7	2

6. $634 - 26$

7. $264 - 168$

8. $524 - 445$

9. $347 - 168$

Brain Builders

Touchdown!

10. 544 people are at a game. 198 people leave after the first quarter. 179 people leave at halftime. How many people are still at the game?

_______ people

11. **Test Practice** Which problem needs to be regrouped twice?

354 − 134 ○

367 − 263 ○

364 − 274 ○

364 − 278 ○

Math at Home Have your child explain how regrouping tens and hundreds are the same.

Name

Lesson 7

Rewrite Three-Digit Subtraction

ESSENTIAL QUESTION
How can I subtract three-digit numbers?

Math in My World

385 − 266

	hundreds	tens	ones
−			

_______ pages

Teacher Directions: Our teacher is reading a very long book to our class. The book has 385 pages. So far, she has read 266 pages of the book. How many pages does she have left to read? Write the numbers in the place-value chart and subtract.

Processes &Practices

Guided Practice

You can rewrite a problem to subtract.
Find 368 − 279.

Step 1 Rewrite.

Step 2 Subtract.

Helpful Hint
Write the greater number at the top. Write the other number below it.

	2	15	18
	~~3~~	~~6~~	~~8~~
−	2	7	9

Line up the ones, tens, and hundreds.

Rewrite the problem. Subtract.

1. 336 − 272

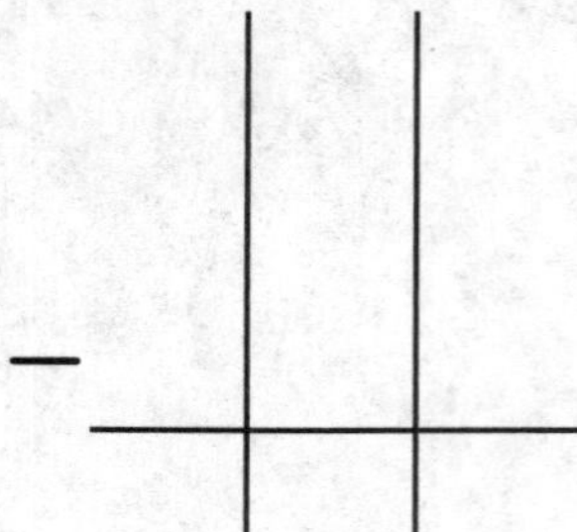

2. 377 − 264

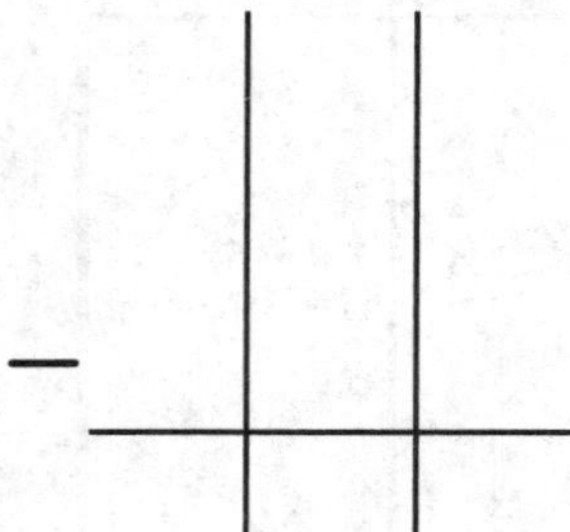

3. 633 − 265

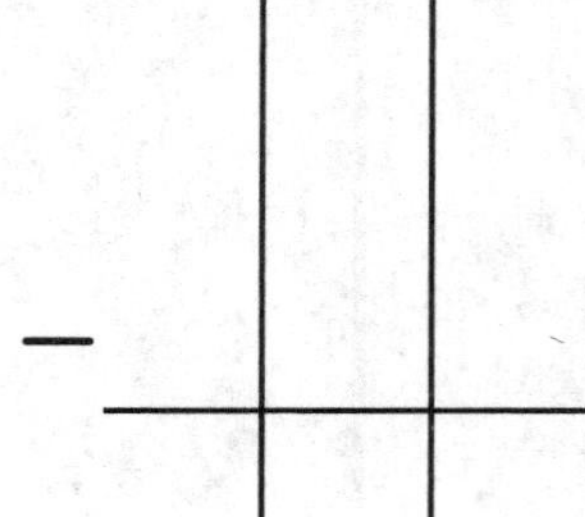

4. 264 − 175

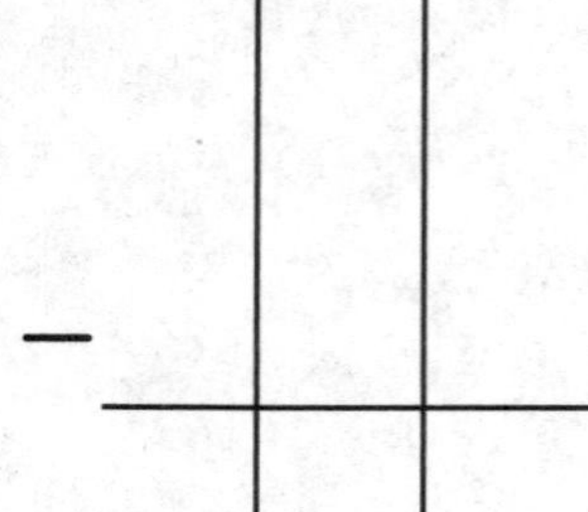

5. 845 − 378

6. 555 − 428

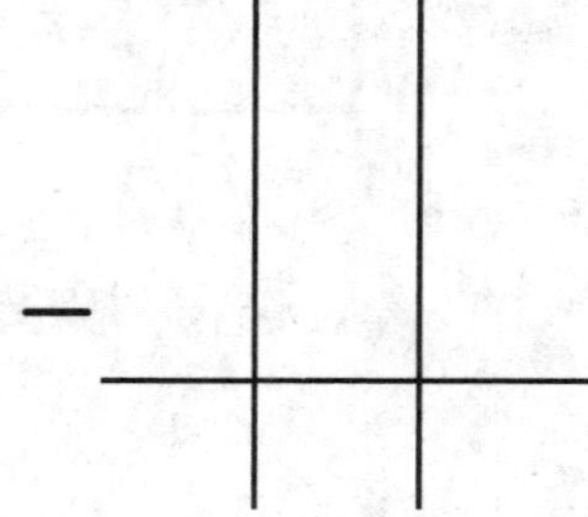

Talk Math How is rewriting three-digit subtraction different than when you rewrite two-digit subtraction?

Name

Independent Practice

Rewrite the problem. Subtract.

7. 363 − 278

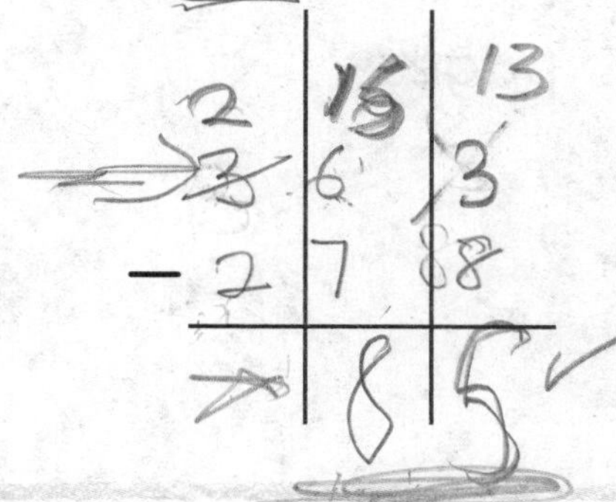

8. 285 − 185

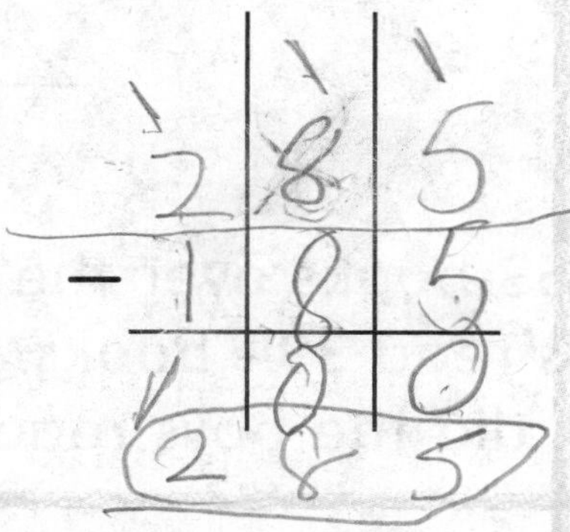

9. 634 − 175

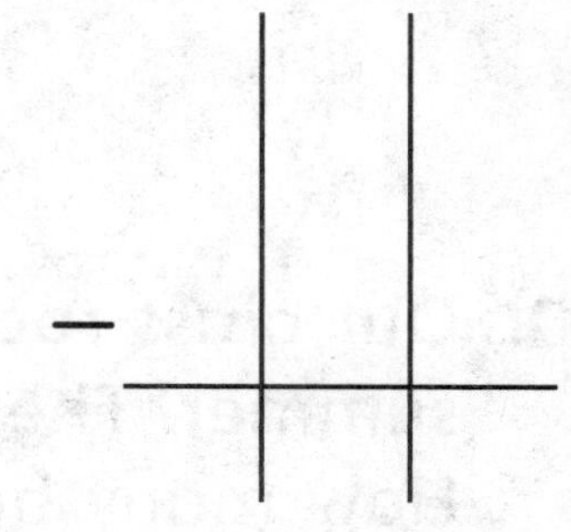

10. 375 − 142

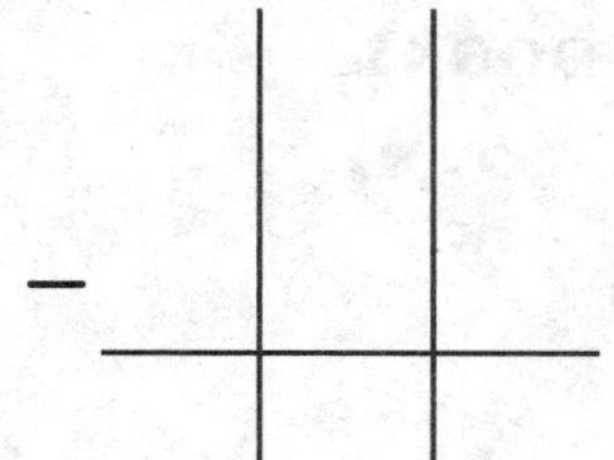

11. 825 − 195

12. 647 − 373

13. 695 − 295

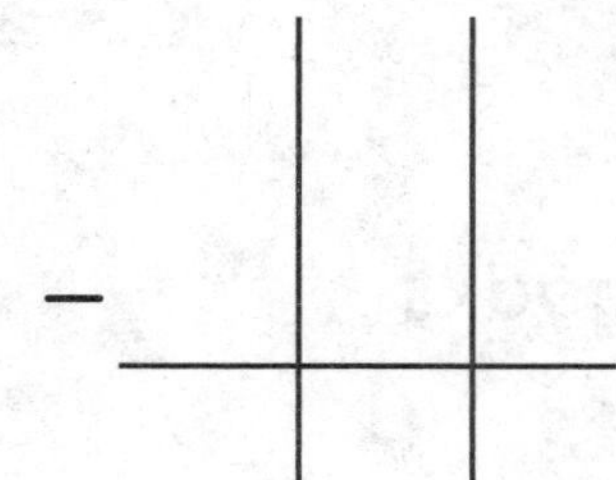

14. 853 − 259

15. 496 − 349

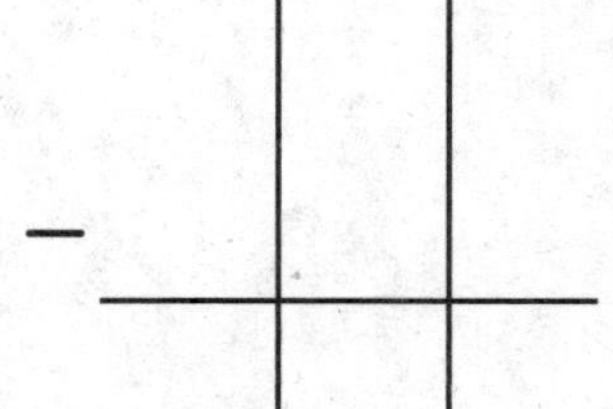

16. 495 − 267

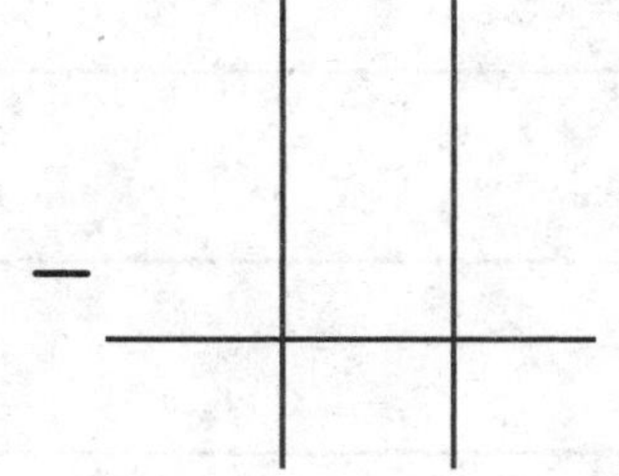

17. 845 − 264

18. 764 − 375

Problem Solving

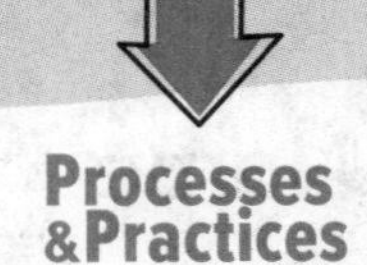

19. Lucas is reading a book that is 239 pages long. He has read 159 pages. How many pages does he have left to read?

_______ pages

20. Our class read 753 books over the summer. The girls read 394 books. How many books did the boys read?

_______ books

Brain Builders

21. Our school library has 125 books about pets and 110 about sports. I have read 96 of the books. How many books do I have left to read?

_______ books

Write Math Explain how you rewrite three-digit subtraction problems to solve.

____

____

____

Name

My Homework

Lesson 7

Rewrite Three-Digit Subtraction

Homework Helper

Need help? connectED.mcgraw-hill.com

Find 356 − 298.

Step 1 Rewrite. Place the greater number on top.

Step 2 Subtract. Regroup if necessary.

	2	14	16
	3	5	6
−	2	9	8
		5	8

Practice

Rewrite the problem. Subtract.

1. 724 − 235

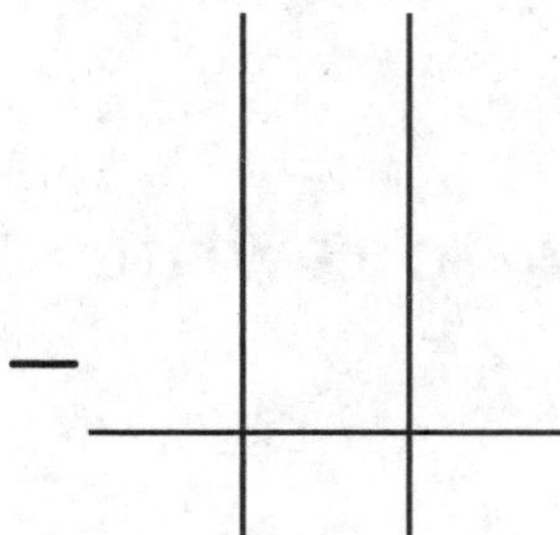

2. 616 − 337

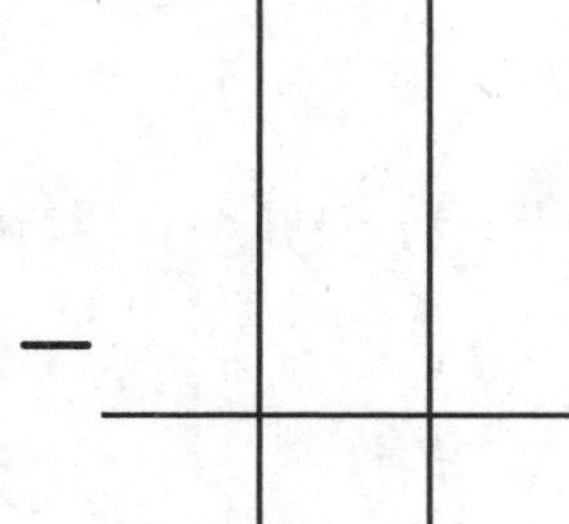

3. 374 − 286

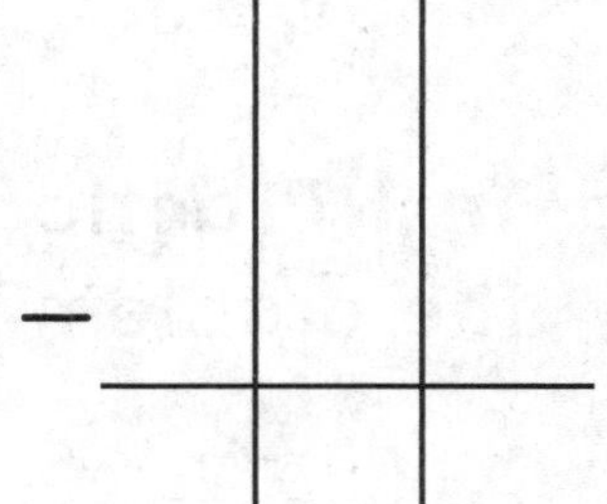

4. 875 − 596

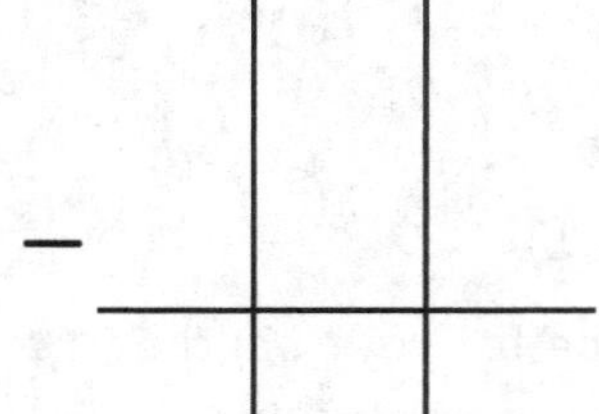

5. 945 − 387

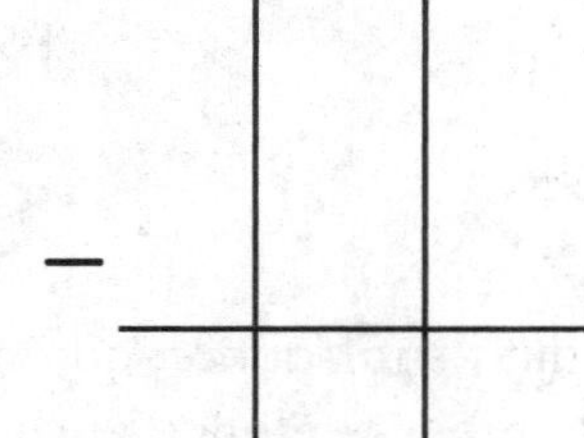

6. 435 − 294

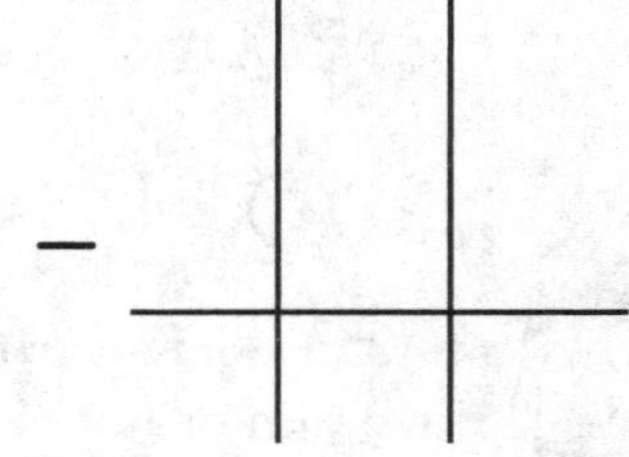

Rewrite the problem. Subtract.

7. 162 − 89

8. 619 − 254

9. 195 − 99

Brain Builders

10. 835 people are on a boat. 295 people get off the boat. Then, 124 more people get on the boat. How many people are on the boat now?

_______ people

11. Test Practice Which problem shows how to rewrite the problem?

368 − 179 = _______

368	179	368	368
− 179	− 368	− 179	+ 179
200	411	189	547
○	○	○	○

Math at Home Write a three-digit subtraction number sentence for your child. Have him or her rewrite the number sentence vertically and then subtract.

Name

Lesson 8

Problem Solving

STRATEGY: Write a Number Sentence

ESSENTIAL QUESTION
How can I subtract three-digit numbers?

Fido always begs for treats. He has a bag of 255 treats. By the end of the month, he has eaten 82 treats. How many treats are left for Fido to eat?

Watch

1 Understand Underline what you know. Circle what you need to find.

2 Plan How will I solve the problem?

3 Solve Write a number sentence.

255 − 82 = 173 treats are left

4 Check Is my answer reasonable? Explain.

Practice the Strategy

385 students entered the science fair. 193 of the students were girls. How many students were boys?

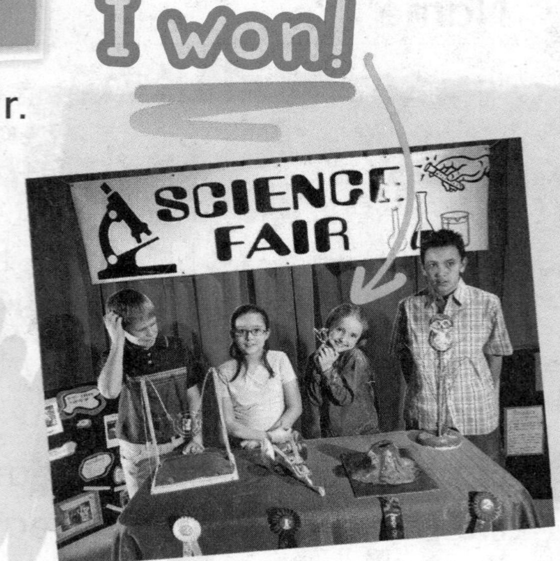

1 Understand Underline what you know.
Circle what you need to find.

2 Plan How will I solve the problem?

3 Solve I will...

______ − ______ = ______

4 Check Is my answer reasonable? Explain.

Name

Apply the Strategy

1. It took 227 days to build railroad tracks over a mountain. It took 132 days to build tracks on flat ground. How many more days did it take to build over the mountain?

2. The Fuller family is driving 475 miles. They have already gone 218 miles. How many miles are left to go?

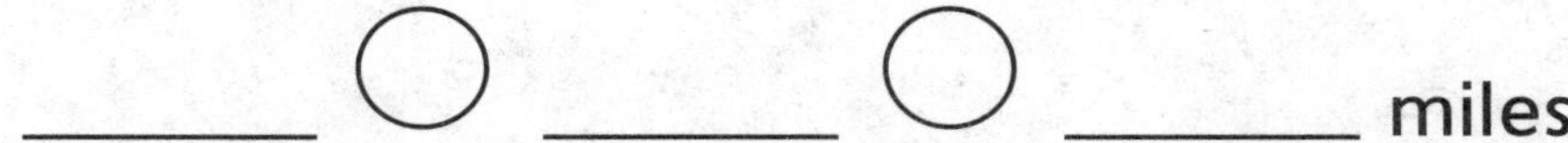

Brain Builders

3. Flora's class is trying to collect 850 cans. Her class turned in 232 cans on Saturday and 138 cans on Sunday. How many cans are still needed to reach the class goal?

_______ cans

Review the Strategies

Choose a strategy

- **Make a model.**
- **Guess, check, and revise.**
- **Write a number sentence.**

4. Anna has been keeping track of the weather for 289 days. 196 of the days were sunny. How many days were not sunny?

_______ days

5. There are 836 flowers in the field. The children pick 398 of the flowers. How many flowers are left in the field?

_______ flowers

6. 423 fish are in the tank. 184 fish are removed from the tank. How many fish are in the tank now?

_______ fish

Name

My Homework

638 people started the race. 459 people finished the race. How many people did not finish the race?

1 Understand Underline what you know. Circle what you need to find.

2 Plan How will I solve the problem?

3 Solve Write a number sentence.

638	–	459	=	179
people started		people finished		people did not finish

179 people did not finish the race.

4 Check Is my answer reasonable?

Underline what you know. Circle what you need to find.

1. A farmer picks 389 ears of corn. He sells 183 ears of corn. How many ears of corn does the farmer have left?

_____ ◯ _____ ◯ _____ ears of corn

Brain Builders

2. 135 people are at the park. 46 people are playing baseball. How many people are not playing baseball? Explain how you solved the problem to a family member.

_____ people

3. There are 276 animals at the zoo. 185 of the animals are male. How many animals are female? How many more animals are male than female?

____

____

4. **Test Practice** 152 tomatoes grew in my garden last year. This year only 98 tomatoes grew. How many more tomatoes grew last year?

252	250	54	52
◯	◯	◯	◯

Math at Home Have your child find 477 – 293 by writing a number sentence.

Name

Lesson 9
Subtract Across Zeros

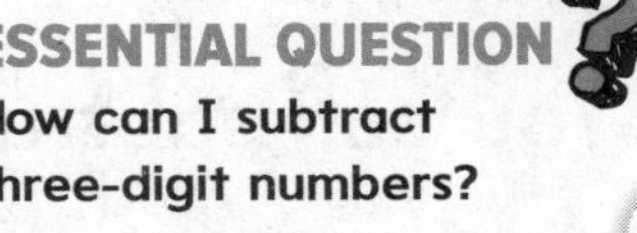
ESSENTIAL QUESTION
How can I subtract three-digit numbers?

Math in My World

_____ animals

Teacher Directions: Model using base-ten blocks. We have learned about 200 different kinds of animals. 126 of those animals were mammals. How many of the animals were not mammals? Draw the blocks. Write the number.

Processes &Practices

Guided Practice

You can subtract across zeros.

Find 400 − 234.

Step 1 Subtract the ones. You cannot subtract 4 ones from 0 ones. There are no tens to subtract from. Look at the hundreds. Regroup 1 hundred as 10 tens. Then regroup 1 ten as ten ones.

	hundreds	tens	ones
	☐	☐	10
	~~4~~	~~0~~	~~0~~
−	2	3	4
			6

Step 2 Subtract the tens. There are 9 tens left. Subtract 3 tens from 9 tens.

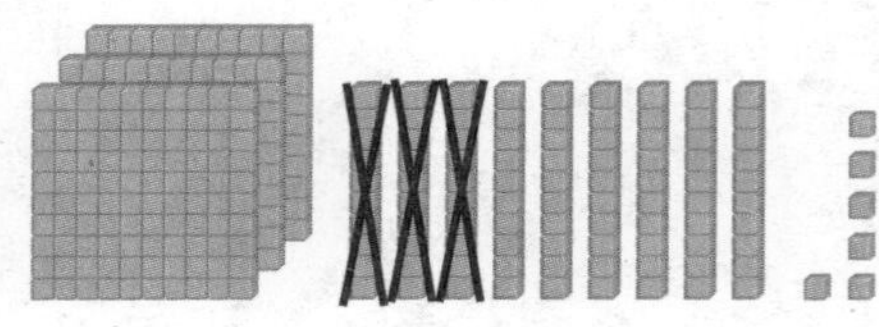

	hundreds	tens	ones
	☐	9	10
	~~4~~	~~0~~	~~0~~
−	2	3	4
		6	6

Step 3 Subtract the hundreds. There are 3 hundreds left. Subtract 2 hundreds from 3 hundreds.

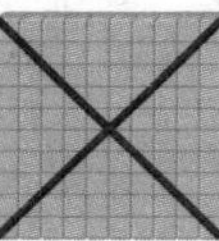

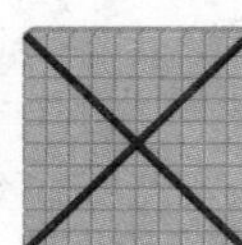

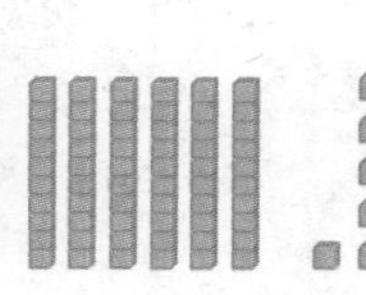

400 − 234 = 166

	hundreds	tens	ones
	3	9	10
	~~4~~	~~0~~	~~0~~
−	2	3	4
	1	6	6

Talk Math How is subtracting from 400 different than subtracting from 435?

Name

Independent Practice

Use Work Mat 7 and base-ten blocks. Subtract.

1.

	hundreds	tens	ones
	□	□	□
	8	0	0
−	5	3	2

2.

	hundreds	tens	ones
	□	□	□
	7	0	0
−	6	1	4

3. $\begin{array}{r} 100 \\ -\ \ 76 \\ \hline \end{array}$

4. $\begin{array}{r} 900 \\ -\ 287 \\ \hline \end{array}$

5. $\begin{array}{r} 400 \\ -\ 167 \\ \hline \end{array}$

6. $\begin{array}{r} 700 \\ -\ 444 \\ \hline \end{array}$

7. $\begin{array}{r} 300 \\ -\ \ 16 \\ \hline \end{array}$

8. $\begin{array}{r} 800 \\ -\ 477 \\ \hline \end{array}$

9. $\begin{array}{r} 900 \\ -\ 876 \\ \hline \end{array}$

10. $\begin{array}{r} 500 \\ -\ \ 54 \\ \hline \end{array}$

11. $\begin{array}{r} 800 \\ -\ 691 \\ \hline \end{array}$

12. $\begin{array}{r} 200 \\ -\ \ 75 \\ \hline \end{array}$

13. $\begin{array}{r} 500 \\ -\ 321 \\ \hline \end{array}$

14. $\begin{array}{r} 600 \\ -\ 312 \\ \hline \end{array}$

Problem Solving

15. 400 bees are in the hive. 145 bees leave the hive. How many bees are still in the hive?

_______ bees

Brain Builders

16. 300 girls will do gymnastics on Saturday. 100 girls did gymnastics on Thursday, and 93 girls did gymnastics on Friday. How many more girls will do gymnastics on Saturday than on Thursday and Friday all together?

_______ girls

17. 500 people signed up for soccer. 123 people stopped playing. Then 154 more people stopped playing. How many people still play soccer? Explain.

____

____

____

Name

My Homework

Lesson 9

Subtract Across Zeros

Homework Helper

Need help? connectED.mcgraw-hill.com

You can subtract across zeros.

Find 600 − 336.

Step 1 Subtract the ones. Regroup 1 hundred into 10 tens. Then regroup 1 ten as 10 ones.

Step 2 Subtract the tens.

Step 3 Subtract the hundreds.

	hundreds	tens	ones
	5	9	10
	~~6~~	~~0~~	~~0~~
−	3	3	6
	2	6	4

Practice

Subtract.

1. $\begin{array}{r} 300 \\ -\ 251 \\ \hline \end{array}$

2. $\begin{array}{r} 600 \\ -\ 139 \\ \hline \end{array}$

3. $\begin{array}{r} 700 \\ -\ 386 \\ \hline \end{array}$

4. $\begin{array}{r} 200 \\ -\ 126 \\ \hline \end{array}$

5. $\begin{array}{r} 500 \\ -\ 385 \\ \hline \end{array}$

6. $\begin{array}{r} 800 \\ -\ 272 \\ \hline \end{array}$

Subtract.

7. $\begin{array}{r} 100 \\ -\ 89 \\ \hline \end{array}$

8. $\begin{array}{r} 600 \\ -\ 564 \\ \hline \end{array}$

9. $\begin{array}{r} 500 \\ -\ 268 \\ \hline \end{array}$

10. $\begin{array}{r} 900 \\ -\ 432 \\ \hline \end{array}$

11. $\begin{array}{r} 700 \\ -\ 364 \\ \hline \end{array}$

12. $\begin{array}{r} 200 \\ -\ 147 \\ \hline \end{array}$

Brain Builders

13. 400 pineapples are planted in a row. 293 pineapples get picked. How many pineapples are still in the row? Explain to a friend or family member how you subtracted across zeros.

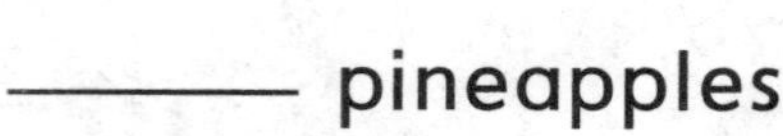

_______ pineapples

14. **Test Practice** Find 600 − 289.

○ 889　○ 311　○ 301　○ 489

Math at Home Have your child explain how to subtract 392 from 800.

Name ....

My Review

Chapter 7
Subtract Three-Digit Numbers

Vocabulary Check 

place value **regroup** **subtract**

Write the correct word in the blank.

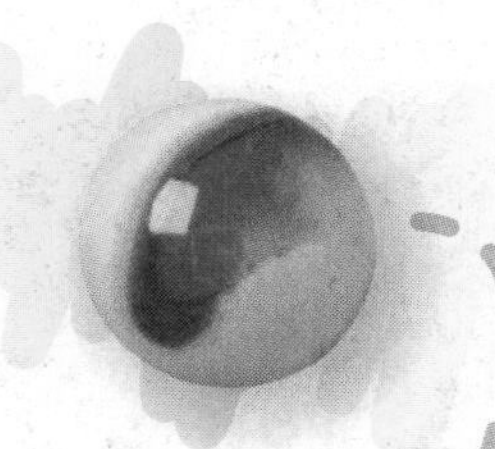

1. To identify the value of the digit 9 in the numbers 976, 93, and 9, use what you know about ______________.

2. You can ______________ to find the difference.

3. To take a number apart and write it a new way is to ______________.

Concept Check

Subtract.

4. 600 − 400 = ______ 5. 700 − 600 = ______

Subtract.

6. $\begin{array}{r} 800 \\ -\ 400 \\ \hline \end{array}$

7. $\begin{array}{r} 900 \\ -\ 600 \\ \hline \end{array}$

8. $\begin{array}{r} 773 \\ -\ 100 \\ \hline \end{array}$

9. $\begin{array}{r} 261 \\ -\ 10 \\ \hline \end{array}$

10. $\begin{array}{r} 938 \\ -\ 329 \\ \hline \end{array}$

11. $\begin{array}{r} 885 \\ -\ 16 \\ \hline \end{array}$

12. $\begin{array}{r} 357 \\ -\ 189 \\ \hline \end{array}$

13. $\begin{array}{r} 987 \\ -\ 598 \\ \hline \end{array}$

14. $\begin{array}{r} 201 \\ -\ 124 \\ \hline \end{array}$

Rewrite the problem. Subtract.

15. 385 − 166

−

16. 247 − 189

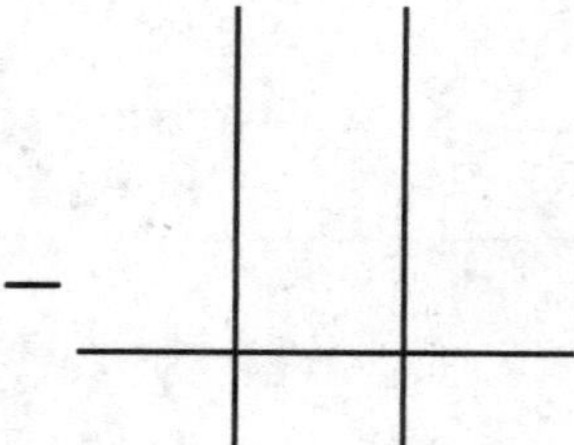

17. 925 − 638

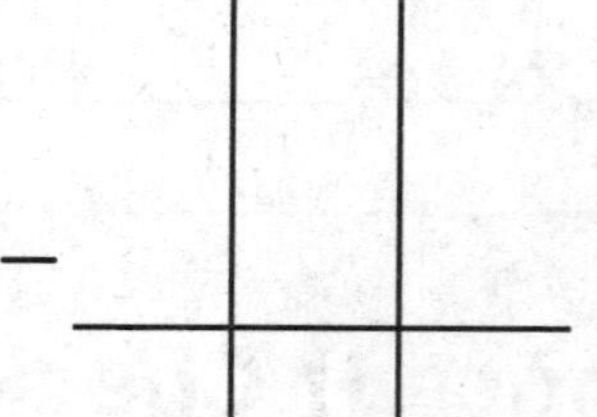

Subtract.

18. $\begin{array}{r} 400 \\ -\ 254 \\ \hline \end{array}$

19. $\begin{array}{r} 700 \\ -\ 443 \\ \hline \end{array}$

20. $\begin{array}{r} 300 \\ -\ 165 \\ \hline \end{array}$

Name

Brain Builders

21. There are 620 buttons. 200 of them are square. 100 are oval. The rest are round. How many buttons are round?

_______ buttons

22. Clara had 359 pennies. She lost some of the pennies. Now she has 266 pennies. How many pennies did Clara lose?

_______ pennies

23. **Test Practice** There are 334 cars in the parking lot on Sunday. On Monday, there are 182 cars. How many more cars were there on Sunday?

334 cars	182 cars	152 cars	150 cars
○	○	○	○

Reflect

Chapter 7

Answering the Essential Question

Show the ways to subtract three-digit numbers.

Subtract hundreds.

800 – 500 = ______

Regroup tens to subtract.

$$\begin{array}{r} 835 \\ -\ 726 \\ \hline \end{array}$$

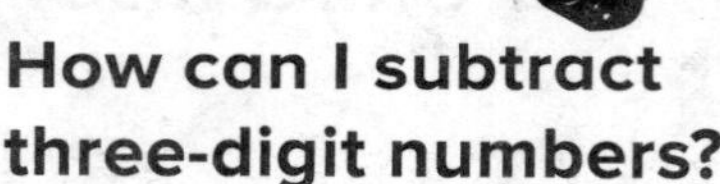

ESSENTIAL QUESTION

How can I subtract three-digit numbers?

Regroup tens and hundreds to subtract.

$$\begin{array}{r} 935 \\ -\ 397 \\ \hline \end{array}$$

Rewrite to subtract.

381 – 298

Name Date

Score

Performance Task

Making Way for the Library

The city is building a new library. To start building, 964 pounds of dirt must be removed. The loader the city used made 3 trips to remove the dirt as shown in the table below.

Load 1	Load 2	Load 3
268 pounds	297 pounds	290 pounds

Show all your work to receive full credit.

Part A

How much dirt was left after removing the first load? Check your answer.

	hundreds	tens	ones
−			

Part B

How many pounds of dirt is left after the second load was removed? Draw base-ten blocks to model the subtraction.

_____________ pounds of dirt

Part C

Did removing the third load get rid of all of the dirt? Explain.

Chapter 8 Money

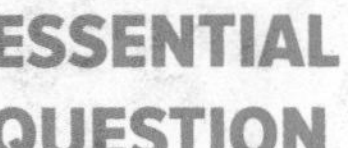

ESSENTIAL QUESTION

How do I count and use money?

Watch a video!

Watch

Name ....

Chapter 8 Project

Classroom Store

1. Each day during the chapter you have earned play money. Today you can use that money to buy items at the Classroom Store!

2. For three of the items you bought, fill in the information below.

I bought ______________________.

It cost ______ ¢.

These are the coins I used to buy it:

I bought ______________________.

It cost ______ ¢.

These are the coins I used to buy it:

I bought ______________________.

It cost ______ ¢.

These are the coins I used to buy it:

Name ....

Am I Ready?

1. Skip count by 5s.

   5, 10, ______, ______, ______, ______

2. Skip count by 10s.

   10, 20, ______, ______, ______, ______

Count on to add.

3. $\begin{array}{r} 25 \\ +\ 1 \\ \hline \end{array}$

4. $\begin{array}{r} 30 \\ +\ 2 \\ \hline \end{array}$

5. $\begin{array}{r} 18 \\ +\ 3 \\ \hline \end{array}$

6. $\begin{array}{r} 16 \\ +\ 4 \\ \hline \end{array}$

7. $\begin{array}{r} 20 \\ +\ 5 \\ \hline \end{array}$

8. $\begin{array}{r} 35 \\ +10 \\ \hline \end{array}$

9. Kim is counting her mittens. She skip counts by 2s. She counts up to 12. How many pairs of mittens does Kim have?

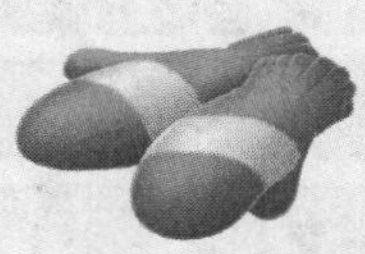

______ pairs

How Did I Do?

Shade the boxes to show the problems you answered correctly.

1	2	3	4	5	6	7	8	9

Name ____________________

My Math Words

Vocab

Review Vocabulary

equal groups   repeated addition   skip count

Look at each example. Complete each sentence to describe the example. Then write the repeated addition sentence.

Example	Describe	Write
	Equal groups of ______. Skip count by ______.	
	Equal groups of ______. Skip count by ______.	
	Equal groups of ______. Skip count by ______.	

My Vocabulary Cards

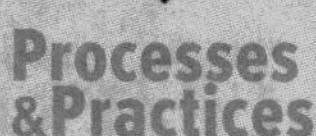

Processes &Practices

Lesson 8-1

cent

→ 1 cent

→ 5 cents

Lesson 8-1

cent sign (¢)

1¢ 5¢

Lesson 8-3

dime

10 cents
10¢

Lesson 8-7

dollar

100 cents

100¢

Lesson 8-7

dollar sign ($)

$1 or $1.00

Lesson 8-2

nickel

5 cents
5¢

Directions:
Ideas for Use

- Have students think of words that rhyme with some of the words.
- Arrange the cards in alphabetical order.
- Tell students to create riddles for each word. Ask them to work with a friend to guess the word for each word card.

The sign used to show cents.

One penny equals one cent or 1¢. 5 pennies equals 5 cents or 5¢.

One dollar has a value of 100 cents or 100¢.

A dime has a value of 10 cents or 10¢.

A nickel has a value of five cents or 5¢.

The sign used to show dollars.

My Vocabulary Cards

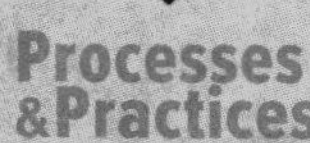

Lesson 8-1

penny

I cent
I¢

Lesson 8-4

quarter

25 cents
25¢

Directions:
Ideas for Use

- Have students use a blank card to write this chapter's essential question. Have them use the back of the card to write or draw examples that help them answer the question.
- Use the blank cards to write your own vocabulary cards.

A quarter has a value of 25 cents or 25¢.

A penny has a value of one cent or 1¢.

My Foldable

FOLDABLES® Follow the steps on the back to make your Foldable.

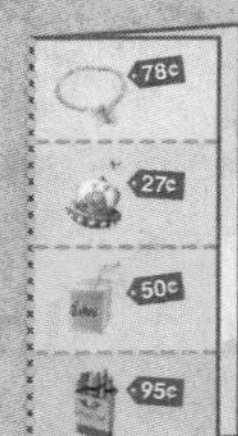

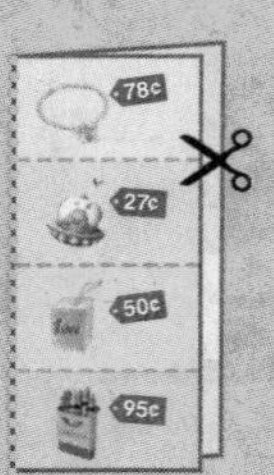

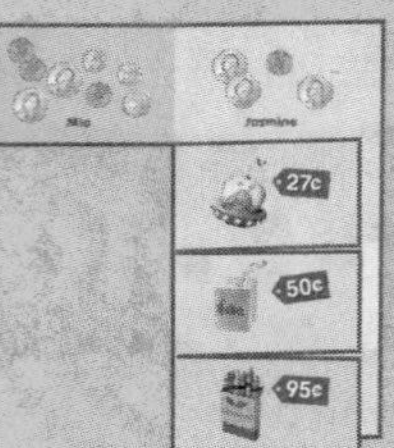

Mia

Jasmine

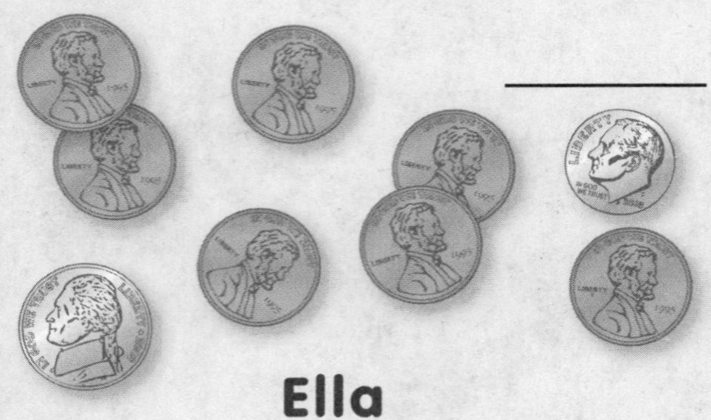

Ella

Ryan

Nathan

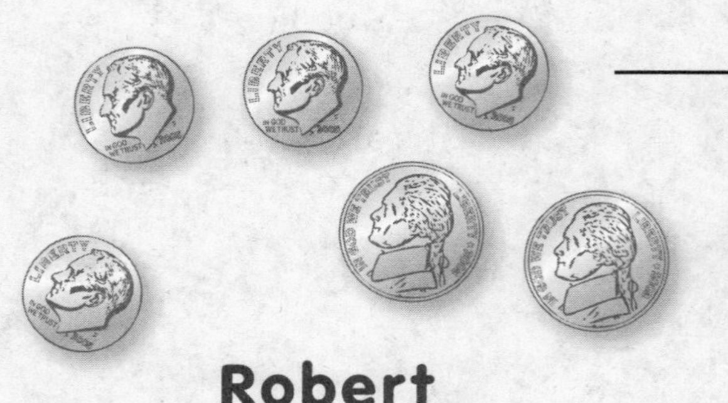

Robert

Name

Lesson 1
Pennies, Nickels, and Dimes

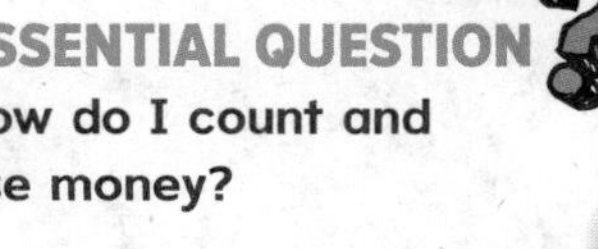

ESSENTIAL QUESTION
How do I count and use money?

Math in My World

pennies
1¢
____

nickels
5¢
____

dimes
10¢
____

Teacher Directions: Use pennies, nickels, and dimes. Sort the coins. Find the value of each group of coins. Write the value on each bouncy ball machine.

Processes & Practices

Guided Practice

Helpful Hint
¢ stands for cents.

dime = 10¢

Count by 10s.

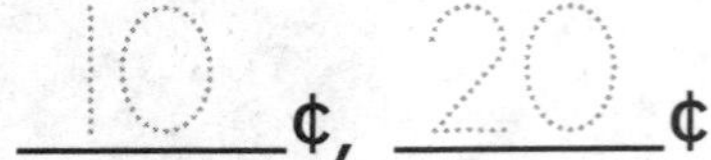

10 ¢, 20 ¢

nickel = 5¢

Count by 5s.

5 ¢, 10 ¢

penny = 1¢

Count by 1s.

1 ¢, 2 ¢

To find the value of coins, start counting with the coin that has the greatest value.

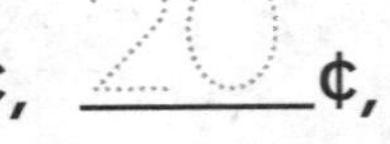

10 ¢, 20 ¢, 25 ¢, 30 ¢, 31 ¢, 32 ¢ = 32 ¢

Count to find the value of the coins.

1.   

   ______¢, ______¢, ______¢ = ______¢

2.      

   ______¢, ______¢, ______¢, ______¢, ______¢, ______¢ = ______¢

Talk Math How many dimes are equal to 70 cents?

Name ______________________

Independent Practice

Count to find the value of the coins.

3.

______¢, ______¢, ______¢, ______¢, ______¢ = ______¢

4.

______¢, ______¢, ______¢, ______¢, ______¢, ______¢, ______¢ = ______¢

5.

______¢, ______¢, ______¢, ______¢, ______¢, ______¢, ______¢ = ______¢

6.

______¢, ______¢, ______¢, ______¢, ______¢, ______¢, ______¢ = ______¢

Problem Solving

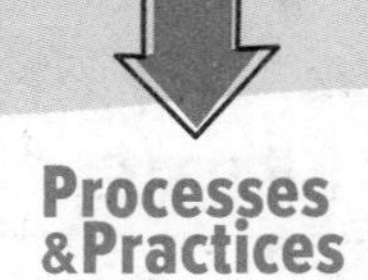

7. Jen had 6 dimes and 4 nickels. She lost 2 of each of them. How much change does she still have?

_______¢

8. Marcy wants to buy beads that cost 80¢ to make a friendship bracelet. If she has 2 nickels, how many dimes does she need to buy the beads?

_______ dimes

Brain Builders

9. Derek has some dimes. He gives Luis 4 dimes. He gives Mia one less dime than he gives to Luis. How much money did Derek give away?

_______¢

10. Paul finds 5 dimes and 2 nickels. He counts them and says he has 50¢. Tell why Paul is wrong. Make it right.

____

____

____

Name

My Homework

Homework Helper

Need help? connectED.mcgraw-hill.com

Helpful Hint
¢ stands for cents.

dime = 10¢

Count by 10s.

10¢, 20¢

nickel = 5¢

Count by 5s.

5¢, 10¢

penny = 1¢

Count by 1s.

1¢, 2¢

To find the value of coins above, start counting with the coin that has the greatest value.

10¢, 20¢, 25¢, 30¢, 31¢, 32¢ = 32¢

Practice

Count to find the value of the coins.

1.     

_____¢, _____¢, _____¢, _____¢, _____¢ = _____¢

2.      

_____¢, _____¢, _____¢, _____¢, _____¢, _____¢ = _____¢

Count to find the value of the coins.

3.      

_____¢, _____¢, _____¢, _____¢, _____¢, _____¢ = _____¢

4.      

_____¢, _____¢, _____¢, _____¢, _____¢, _____¢ = _____¢

Brain Builders

5. Ken has 80¢. His friend has 3 dimes and 10 pennies. How many nickels does his friend need to have the same amount of money as Ken?

_____ nickels

Vocabulary Check 

Circle the correct answer.

6. dime

Math at Home Have your child count coins to total of 90¢.

Name

Lesson 2 Quarters

ESSENTIAL QUESTION
How do I count and use money?

Math in My World

Tools

$

MAGIC MONEY COUNTER

quarters 25¢

dimes 10¢

nickels 5¢

pennies 1¢

Teacher Directions: Use play quarters, dimes, nickels, and pennies. Sort the coins into the correct columns. Count to find the value of each type of coin. Write the value on each column.

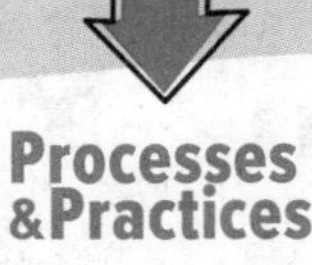

Guided Practice

quarter = 25¢

Count by 25s.

25 ¢, 50 ¢, 75 ¢

Start counting with the coin that has the greatest value.

____¢, ____¢, ____¢, ____¢, ____¢, ____¢, ____¢ = ____¢

Count to find the value of the coins.

1.      

____¢, ____¢, ____¢, ____¢, ____¢, ____¢ = ____¢

2.      

____¢, ____¢, ____¢, ____¢, ____¢, ____¢ = ____¢

Talk Math How many quarters do you need to make 100¢?

Name ______________________

Independent Practice

Count to find the value of the coins.

3.       

_____¢, _____¢, _____¢, _____¢, _____¢, _____¢, _____¢ = _____¢

4.    

_____¢, _____¢, _____¢, _____¢ = _____¢

5.       

_____¢, _____¢, _____¢, _____¢, _____¢, _____¢, _____¢ = _____¢

How many quarters do you need to purchase each item?

6. 

_____ quarters

7. 

_____ quarters

8. 

_____ quarters

Problem Solving

Use the information to answer each question.

9. Dale found a quarter, a dime, and three nickels under the sofa. His mom gave him another quarter. Does he have enough money to buy a school basketball game ticket that costs 50¢?

____

Can Dale also buy a juice box for 25¢?

____

10. Lindsay has 2 quarters and 5 dimes. She gives her friend 1 quarter. Lindsay needs 100¢ to buy a stuffed animal. Does she have enough to buy the toy?

____

Brain Builders

11. Jan has 100¢ in quarters and 10 nickels. She wants to buy bracelets. Each bracelet costs a quarter. How many bracelets can she buy?

_________ bracelets

12. Bryan buys water for 75¢. He uses 3 quarters. Describe another way Bryan could have paid for the water.

____

____

____

Name

My Homework

Homework Helper

Need help? connectED.mcgraw-hill.com

quarter = 25¢

25¢, 50¢, 75¢

Start with the coin that has the greatest value.

25¢, 50¢, 60¢, 70¢, 75¢, 80¢, 81¢ = 81¢

Practice

Count to find the value of the coins.

1.      

_____¢, _____¢, _____¢, _____¢, _____¢, _____¢ = _____¢

2.       

_____¢, _____¢, _____¢, _____¢, _____¢, _____¢, _____¢ = _____¢

Count to find the value of the coins.

3.      

_____ ¢, _____ ¢, _____ ¢, _____ ¢, _____ ¢, _____ ¢ = _____ ¢

Brain Builders

4. Jamal wants to donate 75¢ to the animal shelter. How many quarters would that be? Explain.

____

____

____

5. Jeff has 3 quarters. His friend has 5 dimes. How many more cents does Jeff have than his friend?

_____ ¢

Vocabulary Check

Circle the correct answer.

6. quarter

Math at Home Have your child use quarters to show you 50¢ and 75¢.

Name

Lesson 3
Count Coins

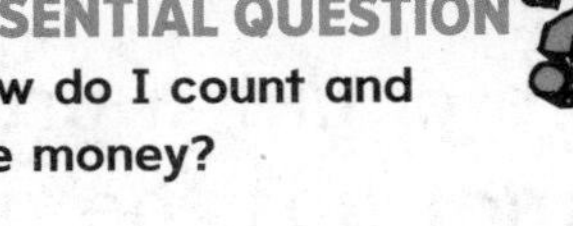

ESSENTIAL QUESTION
How do I count and use money?

Math in My World

Quarters
25¢

Dimes
10¢

Nickels
5¢

Pennies
1¢

The value of all of the coins is ______.

Teacher Directions: Use quarters, dimes, nickels, and pennies. Sort the coins into the appropriate columns. Trace them. Write the total value of the coins.

Processes & Practices

Skip count!

Guided Practice

To count a group of coins, start with the coin that has the greatest value. Count to find the total.

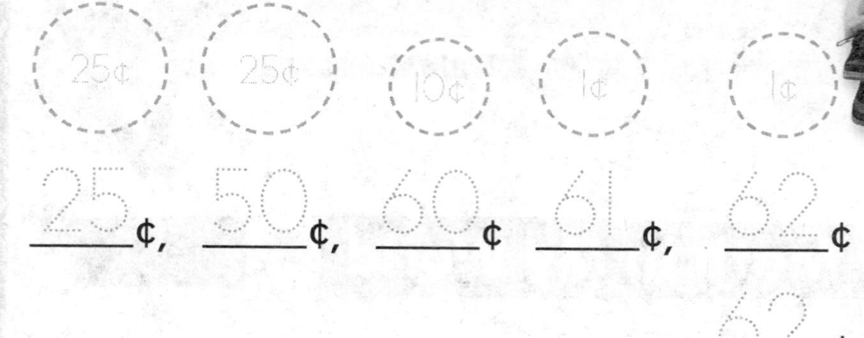

= 62 ¢

Count to find the value of the coins.

1.

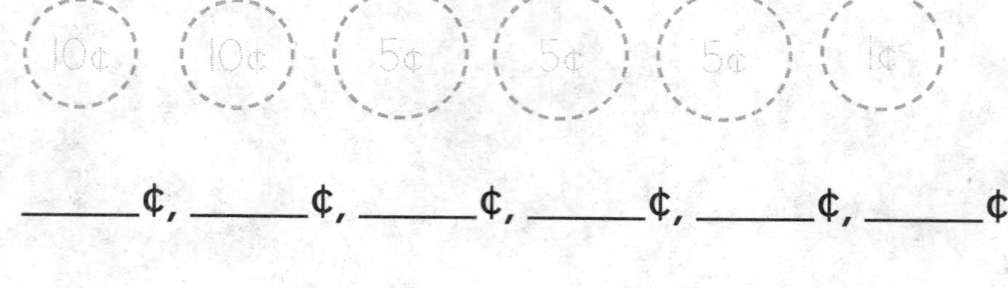

____¢, ____¢, ____¢, ____¢, ____¢, ____¢

= ____¢

2.

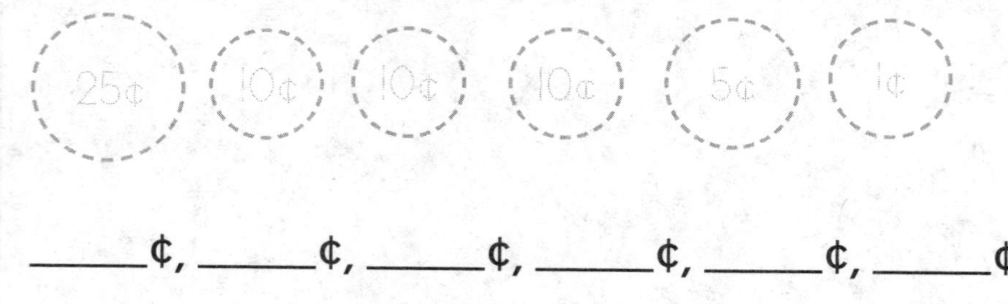

____¢, ____¢, ____¢, ____¢, ____¢, ____¢

= ____¢

Talk Math How does skip counting help you count groups of different coins?

Name

Independent Practice

Count to find the value of the coins.

3. 

_____¢, _____¢, _____¢, _____¢, _____¢, _____¢

= _____¢

4. 

_____¢, _____¢, _____¢, _____¢, _____¢, _____¢

= _____¢

Draw and label the coins from greatest to least. Find the value of the coins.

5. 

= _____¢

Problem Solving

6. Suppose you have 1 quarter, 3 dimes, 1 nickel, and 7 pennies. How much money do you have?

_______¢

7. Luke wants to buy a bouncy ball that costs 25 cents. He has five pennies, 1 dime, and 2 nickels. Does Luke have enough money?

____

Brain Builders

8. Connor has a quarter and a nickel. He gets 2 more quarters for helping around the house. He buys a toy for 25¢. How much money does he have now?

_______¢

Write Math Chase has 5 dimes. Dan has 10 nickels. Who has more money? Explain.

____

____

____

Name

Lesson 3

Count Coins

My Homework

Homework Helper

Need help? connectED.mcgraw-hill.com

To count coins, start with the coin that has the greatest value. Count to find the total value.

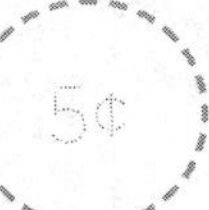

25¢, 50¢, 60¢, 65¢, 70¢, 71¢

= 71¢

Practice

Count to find the value of the coins.

I.

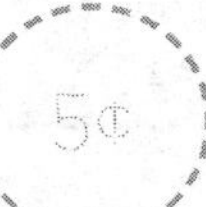
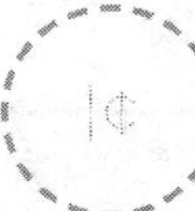

_____ ¢, _____ ¢, _____ ¢, _____ ¢, _____ ¢, _____ ¢

= _____ ¢

2.

_____ ¢, _____ ¢, _____ ¢, _____ ¢, _____ ¢, _____ ¢, _____ ¢

= _____ ¢

Count to find the value of the coins.

3. 

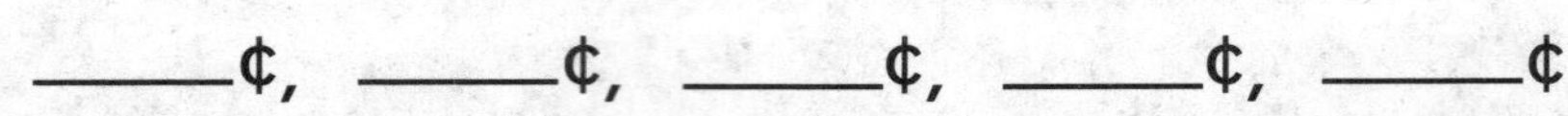

______¢, ______¢, ______¢, ______¢, ______¢

= ______¢

4. 

______¢, ______¢, ______¢, ______¢, ______¢ ______¢

= ______¢

Brain Builders

5. Kate has 6 dimes, 5 nickels, and 4 pennies. She says she has 79¢. Is she correct? Explain.

____

____

____

6. **Test Practice** Which value is more than the value of the coins shown?

41¢	46¢	51¢	36¢
○	○	○	○

Math at Home Give your child coins with a value under $1.00. Have him or her practice counting the coins. Then pretend you are buying and selling things using the coins.

Name ____________________

Check My Progress

Vocabulary Check

penny nickel dime quarter

Complete each sentence.

1. A coin that has a value of 25 cents is a ____________.
2. A coin that has a value of 5 cents is a ____________.
3. A coin that has a value of 1 cent is a ____________.
4. A coin that has a value of 10 cents is a ____________.

Concept Check

Count to find the value of the coins.

5.

_____¢, _____¢, _____¢, _____¢, _____¢, _____¢, _____¢ = _____¢

6.

_____¢, _____¢, _____¢, _____¢, _____¢, _____¢, _____¢ = _____¢

Count to find the value of the coins.

7.      

_____¢, _____¢, _____¢, _____¢, _____¢, _____¢ = _____¢

8.        

_____¢, _____¢, _____¢, _____¢, _____¢, _____¢, _____¢, _____¢

= _____¢

Brain Builders

9. 

Draw a picture to help you solve. What is the value of the coins shown?

= _____¢

10. **Test Practice** Myla needs 55¢ to buy a bag of popcorn. Which coins should she use?

○ quarter, quarter

○ quarter, penny, nickel

○ quarter, quarter, nickel

○ quarter, quarter, penny

Name ______________________________

Lesson 4

Problem Solving

STRATEGY: Act It Out

ESSENTIAL QUESTION

How do I count and use money?

Watch Tools

Gavin has 2 quarters, 1 dime, and 1 nickel. Does he have enough money to buy this toy?

1 Understand

Underline what you know.
Circle what you need to find.

2 Plan

How will I solve the problem?

3 Solve

Act it out.

25¢

25¢, 50¢, 60¢, 65¢ = 65 ¢

Does Gavin have enough money to buy the toy?

yes

4 Check

Is my answer reasonable? Explain.

Practice the Strategy

Robin wants to buy 3 rings.
Each ring costs 20¢. She has
1 quarter, 2 dimes, and 5 nickels.
Does she have enough money?

1 Understand Underline what you know.
Circle what you need to find.

2 Plan How will I solve the problem?

3 Solve I will...

4 Check Is my answer reasonable? Explain.

Name ____________________

Processes &Practices

Apply the Strategy

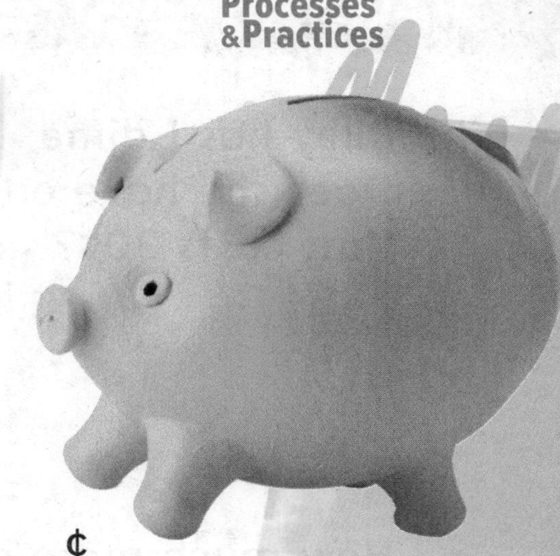

1. Maria has 1 quarter in her piggy bank. Her mom gives her a nickel. Her dad gives her a dime. How much money does Maria have in all?

_______¢

Brain Builders

2. Mark has 2 quarters, 1 dime, and 1 penny. He wants to buy a toy truck for 55¢. Does he have enough money to buy the toy truck? Explain.

____

____

____

3. Wesley has 2 quarters, 3 dimes, and 2 nickels. He has enough money to buy a race car. What is the greatest amount of money that the race car could cost?

_______¢

Brighton has 1 quarter and 6 dimes. Does she have enough money to buy the same car?

____

Review the Strategies

Choose a strategy

- **Act it out.**
- **Draw a picture.**
- **Use logical reasoning.**

4. Riley has 1 dime, 3 nickels, and 4 pennies. Does she have enough to buy a cookie that costs 30¢?

____

How much more does she need?

_______ ¢

5. Annie has coins to buy a gel pen at the store. It costs 85¢. She has 2 quarters and 1 nickel. What two coins does she still need?

____

6. A notebook costs 40¢. What three coins could you use to pay for the toy?

____

Name ....

My Homework

Lesson 4

Problem-Solving: Act It Out

eHelp

Landon has 3 dimes and 2 nickels.
Does he have enough money to buy the pack of stickers that costs 50¢?

1 Understand

Underline what you know.
Circle what you need to find.

2 Plan

How will I solve the problem?

3 Solve

Act it out.

10¢, 20¢, 30¢, 35¢, 40¢ = 40¢

He has 40¢. He needs 50¢.
So, he does not have enough money.

4 Check

Is my answer reasonable?

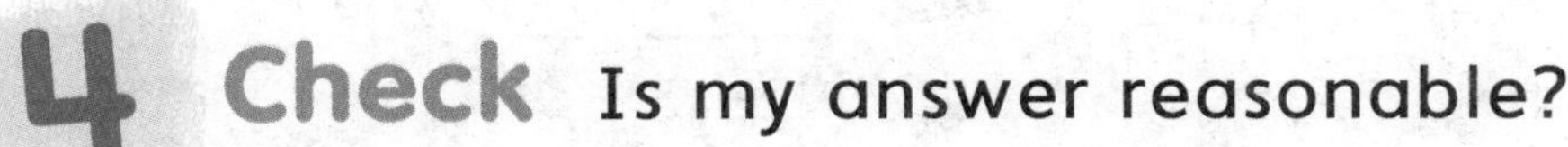

Problem Solving

Underline what you know. Circle what you need to find. Act it out to solve.

1. Jose has 1 quarter, 3 dimes, and 1 nickel. How much more does he need to buy an airplane that costs 75¢?

_______ ¢

2. Leah has 1 quarter, 2 dimes, and a nickel. Her sister has 3 nickels. How many more cents do they need to have 90¢?

_______ ¢

Brain Builders

3. Cole has 75 cents. He has 2 quarters and 1 dime. If the rest of his coins are nickels, how many nickels does he have? Explain how you know to a friend or family member.

_______ nickels

4. Sarah has 1 quarter, 1 dime, 3 nickels, and 3 pennies. Bao has 4 dimes, 2 nickels, and 2 pennies. Who has more money?

____

Math at Home Have your child show you the coins needed to buy a toy that costs 64¢.

Name

Lesson 5
Dollars

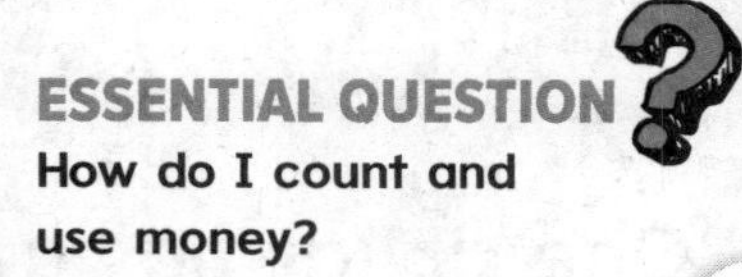

ESSENTIAL QUESTION

How do I count and use money?

Math in My World

one dollar = 100 cents

______ ______ ______ ______

Teacher Directions: Count pennies to 100¢. Write the number of quarters it takes to equal 100¢. Do the same for the number of dimes and the number of nickels.

Processes & Practices

Dollar Sign $1.00 Decimal Point

Guided Practice

One **dollar** has a value of 100 cents or 100¢.
To write one dollar, use a **dollar sign.**

Use a decimal point to separate the dollars from the cents.

one dollar bill = $1.00

100 pennies = $1	20 nickels = $1	10 dimes = $1	4 quarters = $1

Count to find the value of the coins.
Circle the combinations that equal $1.00.

1.

______

2.

______

Talk Math How are $ and ¢ different? How are they alike?

Name

Independent Practice

Count to find the value of the coins. Circle the combinations that equal $1.00.

Helpful Hint
Use a dollar sign to write dollars. Use a cent sign to write cents.

3.

4.

5.

6.

7.

8.

Problem Solving

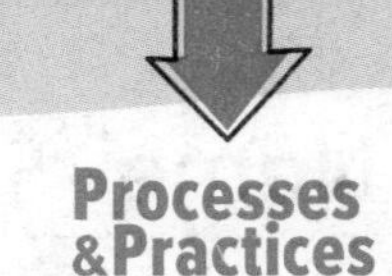

9. Natasha has 1 quarter, 2 dimes, 10 nickels, and 4 pennies. She needs 1 dollar to buy a joke book. How much money does she have?

____

How much more does she need to have one dollar?

____

Brain Builders

10. Chip needs 1 dollar. He has three quarters and one dime. How much does he have?

____

How many more nickels would he need to make 1 dollar?

_______ nickels

Write Math Think of 2 combinations of coins that equal one dollar and write them here.

____

____

____

Name ....

My Homework

Lesson 5

Dollars

Homework Helper

Need help? connectED.mcgraw-hill.com

One dollar has a value of 100 cents or 100¢. To write 1 dollar, use a dollar sign.

one dollar = $1.00

$1.00 is equal to

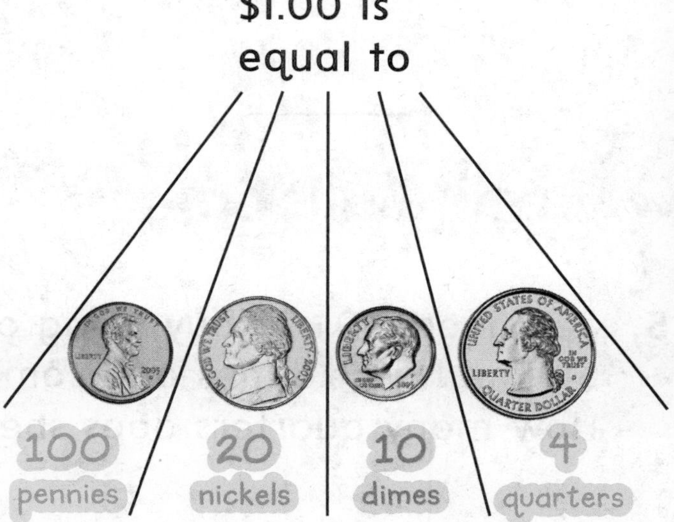

Practice

Count to find the value of the coins. Circle combinations that equal $1.00.

1.

______

2.

______

Count the coins. Write the value.
Circle combinations that equal $1.00.

3. 

____

4. 

____

Brain Builders

5. Jen needs 10¢ to buy 1 bag of pretzels for $1. She has 4 dimes and some quarters. How many quarters does she have?

_______ quarters

6. Diego has 3 quarters, 1 dime, and 1 nickel. How many more nickels does he need to have $1?

_______ nickels

Explain how you know to a friend or family member.

Vocabulary Check 

Circle the correct choices.

7. **one dollar** $1 1$ $1.00 1¢

Math at Home Have your child use various coins to show you two ways to make $1.

Name

My Review

Chapter 8

Answering the Essential Question

Vocabulary Check

Draw lines to match.

1. dime	1 cent or 1¢
2. penny	25 cents or 25¢
3. quarter	5 cents or 5¢
4. dollar	10 cents or 10¢
5. nickel	100 cents or 100¢

Concept Check

Count to find the value of the coins.

6.     

_____¢, _____¢, _____¢, _____¢, _____¢ = _____ ¢

Concept Check

Count to find the value of the coins.

7.      

_____¢, _____¢, _____¢, _____¢, _____¢, _____¢ = _____¢

Count to find the value of the group of coins.

8. 

25¢ 25¢ 10¢ 10¢ 5¢ 5¢

_____¢, _____¢, _____¢, _____¢, _____¢, _____¢

= _____¢

Count to find the value of the coins.
Circle the combinations that equal $1.00.

9. 

_____

10. 

_____

Use each coin to make one dollar.
Write the number of coins you used.

11. 

_____

12. 

_____

Name ....

Problem Solving

13. Lupe buys a toy dinosaur for 47¢. He gives the cashier 1 quarter and 1 dime. How much more money does he need to give the cashier?

____

Circle the three coins Lupe should give the cashier.

Brain Builders

14. John and Mark want to buy a toy that costs 1 dollar. John has 2 dimes and Mark has 2 quarters. How much more money do the boys need? ________

15. **Test Practice** Lacey found 1 quarter and 1 dime. She already had 30¢. Kyra has 85¢. How much more money does Kyra have?

10¢	20¢	25¢	30¢
○	○	○	○

Reflect

Chapter 8

Answering the Essential Question

I cent or I¢

_____, _____, _____ = _____

_____ pennies make $1.00.

5 cents or 5¢

_____, _____, _____ = _____

_____ nickels make $1.00.

ESSENTIAL QUESTION

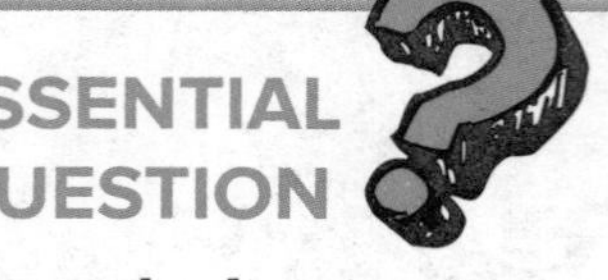

How do I count and use money?

10 cents or 10¢

_____, _____, _____ = _____

_____ dimes make $1.00.

25 cents or 25¢

_____, _____, _____ = _____

_____ quarters make $1.00.

Success is yours!

Name

Date

Score

Performance Task

Brain Builders

Art Supplies

Discount Art Store is having a sale on art supplies. The table below shows the prices for some sale items at the store.

Item	Price
Colored paper	4 for $1
Ruler	30¢
Glue stick	45¢
Marker	85¢

Show all your work to receive full credit.

Part A

Jason bought some art supplies. He gave the cashier exact change. The coins he gave her looked like this:

What items did Jason buy? Explain.

Part B

Olivia wants to buy 2 glue sticks or 2 pieces of colored paper. Which items cost more? Explain.

Part C

Olivia decided to buy 2 pieces of colored paper. How much more or less did she spend than Jason spent in **Part A**? Show the difference using coins.

Part D

Suppose Jason and Olivia each had one dollar to spend at the Discount Art Store. Do either of them have enough money left to buy a ruler after their purchases? Explain.

Chapter 9

Data Analysis

ESSENTIAL QUESTION
How can I record and analyze data?

Our Bodies Need Healthful Food!

Watch a video!

Watch

Name

Chapter 9 Project

Graph Data

1. Use your topic to create a survey question.
2. Ask at least 15 people your survey question.
3. Decide which type of graph would show your data best.
4. Create your graph below.
5. Write a few sentences about what your graph shows.

Create your graph below.

My graph shows:

Name

Am I Ready?

Circle the group that has more.

1.  

Use the tally chart to answer the questions.

Favorite Pet	Tally
Cat	|||
Hamster	||||
Dog	~~||||~~ ||

2. What pet has 4 tally marks?

____

3. How many people like cats?

_____ people

Use the picture to solve.

4. Roy walks four dogs every day. How many brown dogs does he walk?

_____ brown dogs

How Did I Do?

Shade the boxes to show the problems you answered correctly.

1	2	3	4

Name

My Math Words

Review Vocabulary

compare graph tally marks

Use the review words to describe each example.

What I Can Show on a Graph

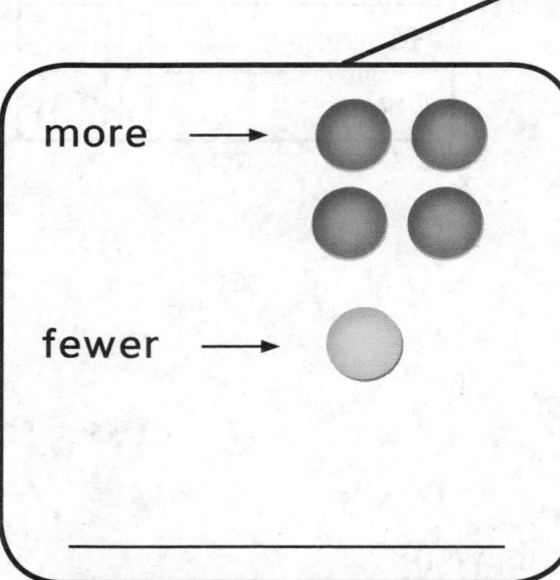

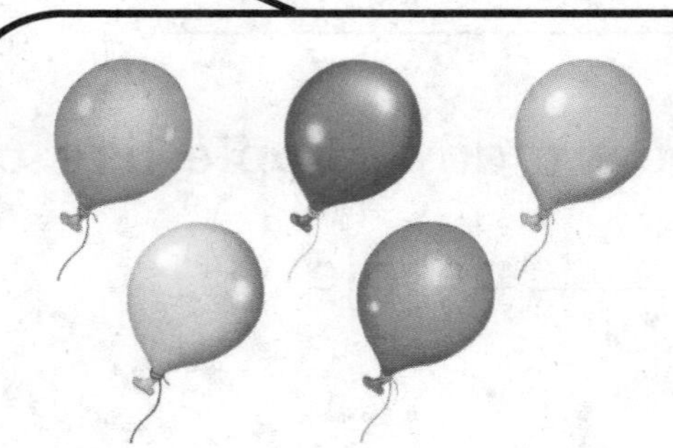

How many balloons?

What is a graph?

My Vocabulary Cards

Lesson 9-4

bar graph

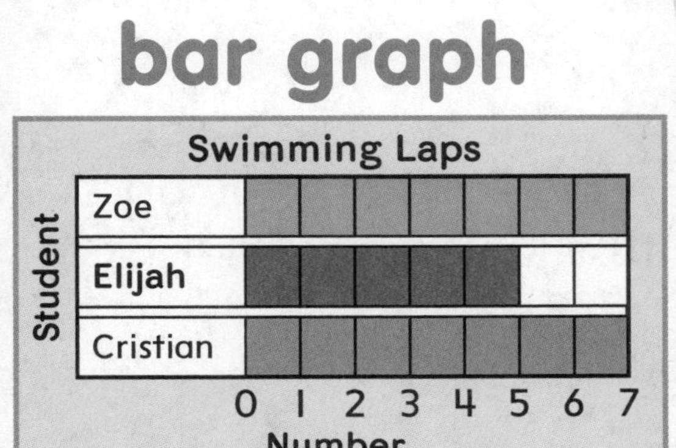

Lesson 9-1

data

What Pet Do You Like Best?	
Pet	Tally
Dogs	𝍸 𝍸 I
Cats	𝍸 II

Lesson 9-2

key

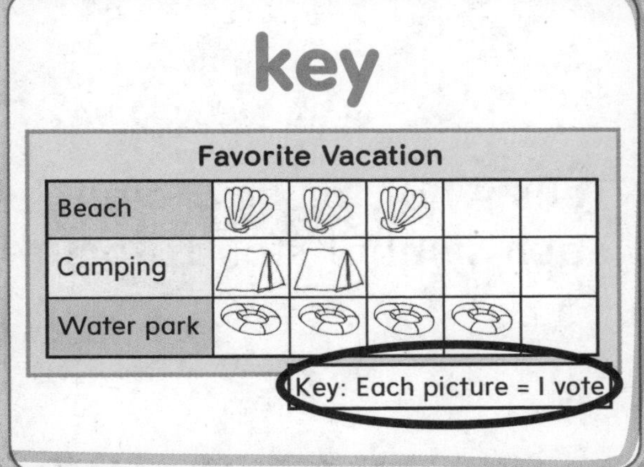

Lesson 9-7

line plot

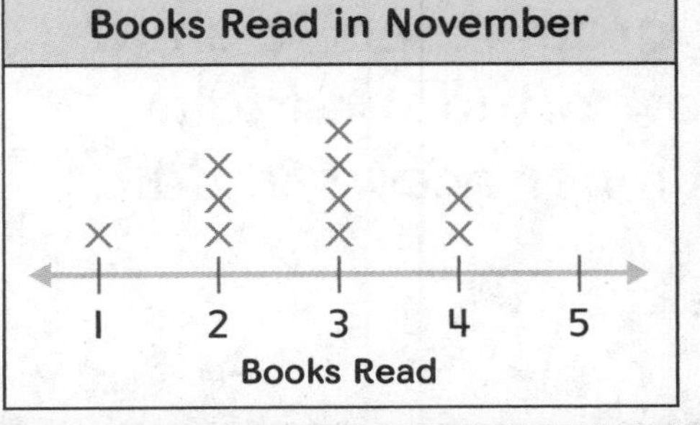

Lesson 9-2

picture graph

Lesson 9-1

survey

Favorite Time of Day	Tally	Total
Morning	𝍸 I	6
Afternoon	III	3
Evening	𝍸 III	8

Teacher Directions:
Ideas for Use

- Have students group 2 or 3 common words. Ask them to add a word that is unrelated to the group. Have them work with a friend to name the unrelated word.
- Have students write a tally mark on each card every time they read the word in this chapter or use it in their writing.

Numbers or symbols that show information

A graph that uses bars to show data

A graph used to show how often a certain number occurs in data

Tells what, or how many, each symbol stands for

To collect data by asking people the same question

A graph that has different pictures to show information collected

My Vocabulary Cards

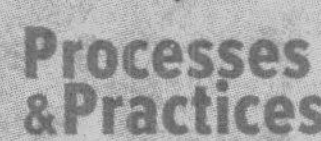

Lesson 9-2

symbol

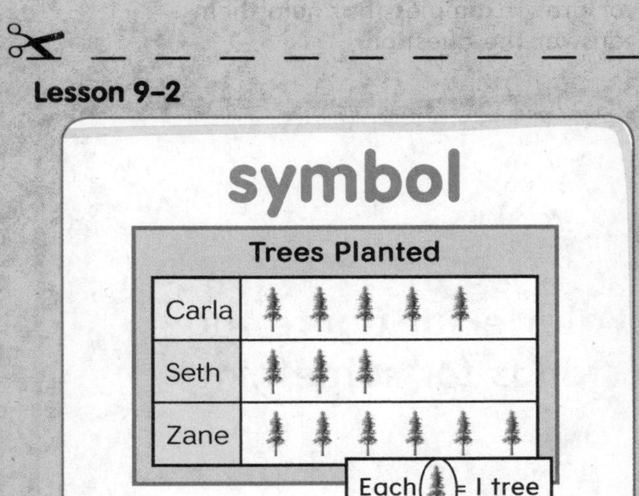

Lesson 9-1

tally marks

Items Sold at School Store	
Item	Tally
Eraser	卌
Bottle of glue	卌 卌
Pencil	卌 |||
Scissors	||

Teacher Directions:
More Ideas for Use

- Ask students to group like ideas they find throughout the chapter, such as using different ways to display data. Have students share the strategies they use to understand the concepts.
- Have students use a blank card to write this chapter's Essential Question. Have them use the back of the card to write or draw examples that help them answer the question.

Marks used in a survey to collect data

A letter or figure that stands for something

My Foldable

FOLDABLES® Follow the steps on the back to make your Foldable.

FOLDABLES®
Study Organizer
1
2

Name

Lesson 1
Take a Survey

ESSENTIAL QUESTION
How can I record and analyze data?

Math in My World

What is your favorite fruit?		
Fruit	Tally	Total
Apple		____
Orange		____
Grape		____

Teacher Directions: Ask 10 classmates to name their favorite fruit. Mark a tally for each choice. Write the total for each fruit. Which fruit do most people like?

Guided Practice

When you take a **survey**, you ask a question. Use **tally marks** to record the answers, or data. **Data** is information.

Taking a survey about books

1. Write a question.
2. Ask your question.
3. Record each person's answer with a tally mark.
4. Count the tally marks.

Type of Book	Tally	Total
Scary	||	2
Funny	𝍸 |||	8
Sports	𝍸	5

Ask 10 students their favorite season. Use tally marks to record the data.

Use the data in the chart to answer the questions.

Favorite Season	Tally	Total
Winter		
Spring		
Summer		
Fall		

1. What season do students like best?

   ____

2. How many students like winter and fall? Write a number sentence to solve.

   _____ ◯ _____ = _____

Talk Math How do tally marks help you organize data?

Name

Independent Practice

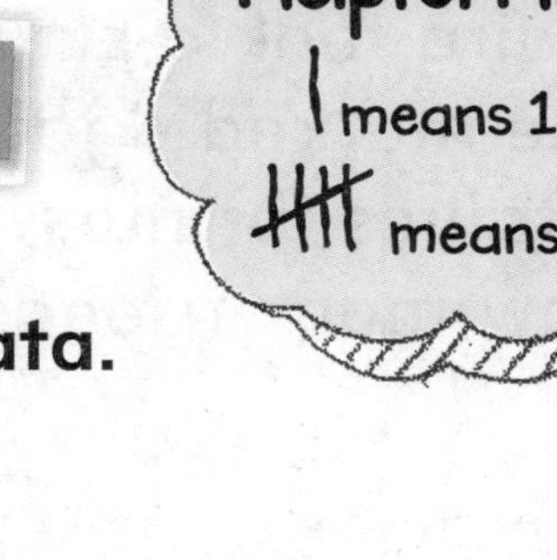

Ask 10 students their favorite picnic food. Use tally marks to record the data.

Picnic Food	Tally	Total
Hamburger		
Hot dog		
Chicken		

Use the data in the chart to answer the questions.

3. How many students like hamburgers? ________

4. How many students like hot dogs? ________

5. What is the favorite picnic food?

____

6. What is the least favorite picnic food?

____

7. How would the chart change if you added your favorite food? ______________________________

____

Problem Solving

8. Claire took a survey of favorite snacks. She marked 6 tallies for vegetables, 4 tallies for nuts, and 8 tallies for fruit. How many friends did she survey?

_______ friends

Brain Builders

9. Students took a survey for Fun Friday activities. Seven students want a picnic. Five students want to go to the zoo. Eight students want to watch a movie. Which chart shows these results? Circle the correct chart.

Activity	Tally
Picnic	𝍸 I
Zoo	𝍸
Movie	III

Activity	Tally
Picnic	𝍸 II
Zoo	𝍸
Movie	𝍸 III

How many students were surveyed? Explain to a friend how you know.

_______ students

10. Why do you think the fifth tally mark is shown at a slant?

____

____

____

Name ....

My Homework

Homework Helper

Need help? connectED.mcgraw-hill.com

A tally chart shows the results of a survey.

Favorite Vegetable	Tally	Total
Carrot	𝍸 I	6
Broccoli	III	3
Corn	𝍸 III	8

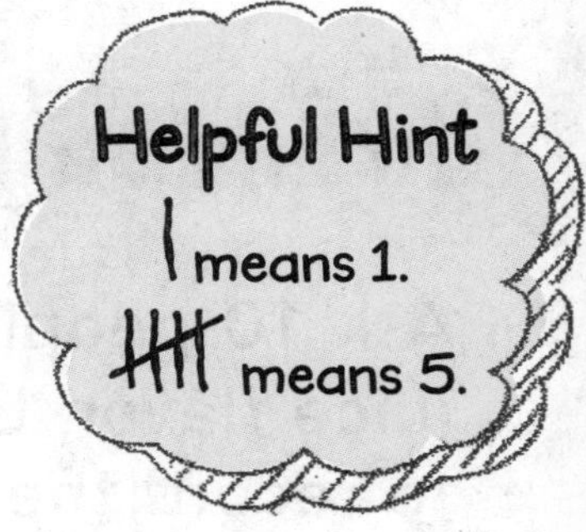

Practice

Ask 10 people their favorite exercise. Use tally marks to record the data.

Use the data in the chart to answer the questions.

Favorite Exercise	Tally	Total
Jump Rope		
Run		
Dance		
Play Baseball		

1. How many people like to run and dance? ______________

2. Do more people like to run and play baseball or dance and jump rope? ______________

Ask 10 people their favorite snack. Use tally marks to record the data.

Use the data in the chart to answer the questions.

Favorite Snack	Tally	Total
Fruit		
Chips		
Cookies		
Carrot Sticks		

3. What snack do people like the most? ____________

4. How many people like healthful snacks the most? ____________

Brain Builders

5. Ask 10 people their favorite juice flavor. Use tally marks to record the data.

Use the data in the chart to answer the questions.

Favorite Juice	Tally	Total
Apple		
Orange		
Cranberry		
Tomato		

6. Write the juices in order from least liked to most liked.

____

Vocabulary Check 

Circle the word that matches the definition.

7. Numbers or symbols that show information

tally marks **survey** **data**

Math at Home Help your child create a survey that they can give to family members.

Name

Lesson 2
Make Picture Graphs

ESSENTIAL QUESTION

How can I record and analyze data?

Math in My World

Favorite Lunch Food

Sandwich					
Pizza					
Hot Dog					
Soup					

Key: Each food = 1 vote

Teacher Directions: Ask five classmates to pick their favorite lunch food. Draw a picture in the chart to show each person's choice.

Guided Practice

You show data with a **picture graph**.
The pictures are a **symbol** for the data.

Favorite Marble Color

Blue	◌	◌	◌		
Red	◌	◌	◌		

Key: Each marble = 1 vote

Helpful Hint
The key tells how many each symbol stands for.

Use the tally chart to make a picture graph.

1.

Favorite Shoes	Tally	Total
Tennis Shoes	|||	3
Flip Flops	~~||||~~	5
Boots	|	1

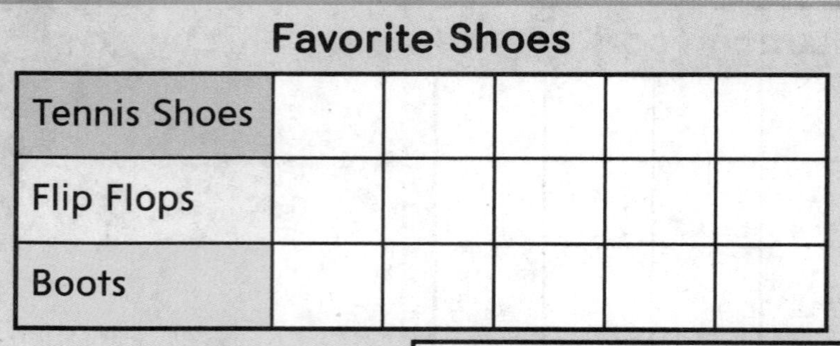

Favorite Shoes

Tennis Shoes					
Flip Flops					
Boots					

Key: Each shoe = 1 vote

Talk Math How are picture graphs different from tally charts?

Name

Independent Practice

Use each tally chart to make a picture graph.

2.

Favorite Drink	Tally	Total
Milk	|||	3
Apple Juice	||	2
Water	||||	4
Lemonade	||||	4

Favorite Drink

Milk					
Apple Juice					
Water					
Lemonade					

Key: Each drink = 1 vote

3.

Favorite Sport	Tally	Total
Baseball	|||	3
Basketball	||||	4
Soccer	||	2
Football	𝍸	5

Favorite Sport

Baseball					
Basketball					
Soccer					
Football					

Key: Each ball = 1 vote

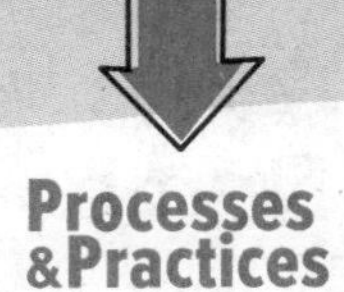

Processes & Practices

Brain Builders

Use the information to make a picture graph.

4. Lilian asked ten people their favorite flower. One said tulips. The same number of people said daisies and carnations. 5 said roses.

Favorite Flower

Tulip					
Daisy					
Rose					
Carnation					

Key: Each flower = 1 vote

5. A school asked 15 students their favorite subject. 3 students said science and 3 said reading. One less person said art than math.

Favorite Subject

Math					
Science					
Reading					
Art					

Key: Each book = 1 vote

Write Math Explain how picture graphs can be more helpful than tally charts.

Name ....

Lesson 2

Make Picture Graphs

My Homework

Homework Helper

Need help? connectED.mcgraw-hill.com

You can use data from a tally chart to make a picture graph.

Favorite Milk	Tally	Total
Chocolate	卌	5
White	II	2
Strawberry	III	3

Favorite Milk

Chocolate					
White					
Strawberry					

Key: Each glass = 1 vote

Practice

Use the tally chart to make a picture graph.

1.

Favorite Animal	Tally	Total
Elephant	IIII	4
Giraffe	卌	5
Bear	IIII	4
Snake	III	3

Favorite Animal

Elephant					
Giraffe					
Bear					
Snake					

Key: Each picture = 1 vote

Brain Builders

Use the tally chart to make a picture graph.

2.

Favorite Cookie	Tally	Total
Chocolate Chip	||||	4
Peanut Butter	卌	5
Oatmeal Raisin	||||	4
Sugar	|||	3

Favorite Cookie

Chocolate Chip					
Peanut Butter					
Oatmeal Raisin					
Sugar					

Key: Each cookie = 1 vote

Explain to a friend or family member how you made the picture graph.

Vocabulary Check

Circle the picture that matches the word.

3. **picture graph**

Type of Book	Tally	Total
Scary	||	
Funny	卌 |||	
Sports	卌	

Favorite Pet

Fish	🐟	🐟	🐟	🐟	🐟	🐟
Dog	🐶	🐶				
Cat	🐱	🐱	🐱	🐱		

Key: Each animal picture = 1 vote

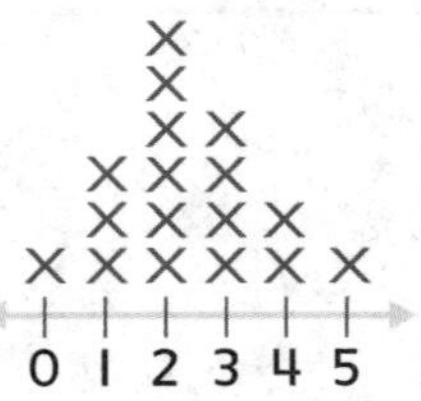

Math at Home Construct a large picture graph. Have your family graph their favorite dinner.

Name ....

Lesson 3

Analyze Picture Graphs

ESSENTIAL QUESTION

How can I record and analyze data?

Math in My World

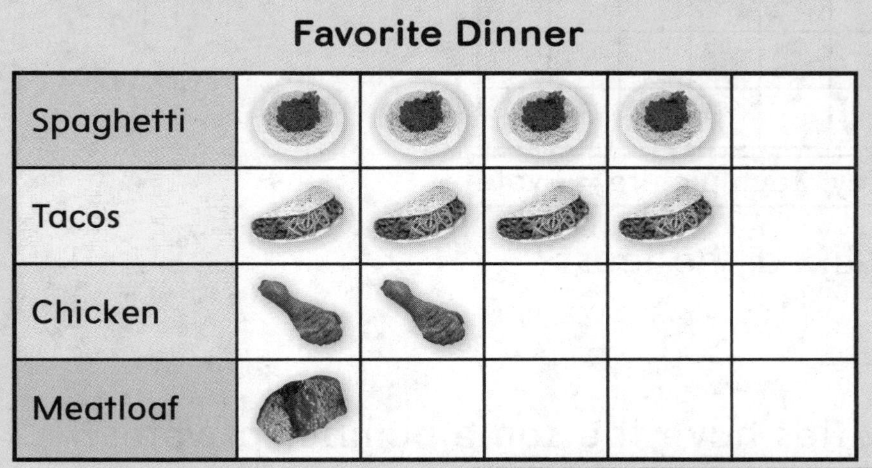

Favorite Dinner

Spaghetti					
Tacos					
Chicken					
Meatloaf					

Key: Each food = 1 vote

Which dinner is the least favorite? ____________________

Teacher Directions: Use the data in the picture graph. Write the number of votes for each dinner. Write the least favorite dinner.

Processes
&Practices

Guided Practice

You can use a picture graph to answer questions.

Favorite Pet

Fish						
Dog						
Cat						

Key: Each animal picture = 1 vote

What pet is the favorite?

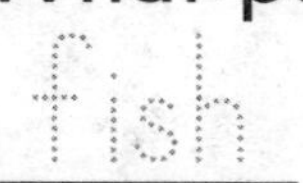

How many votes does each picture show? 1

Use the data from the graph to answer the questions.

Favorite Summer Activity

Swim					
Bike Ride					
Inline Skate					
Play Baseball					

Key: Each picture = 1 vote

1. What activity is liked the least?

   ____

2. Which two activities have the same number of votes?

   ____

3. How many people like to swim and play baseball in all?

   ____

How would you count the votes for bike riding if each picture stood for 2 votes?

Name

Independent Practice

Use the data from the graph to answer the questions.

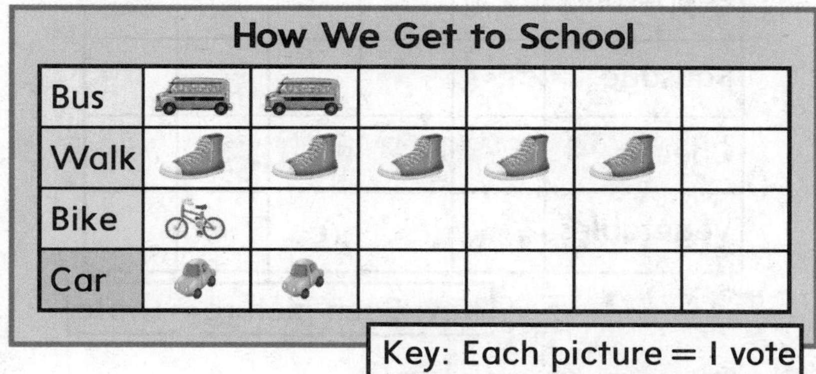

4. How do most students get to school? ____________

5. How many students ride the bus and ride in cars? ______

6. How many students voted? ______

7. How will the graph change if you add another way to get to school?

____

8. Write a question that can be answered using the data in the picture graph.

____

____

9. Write a the number sentence you can use to answer your question.

_____ ◯ _____ ◯ _____

Problem Solving

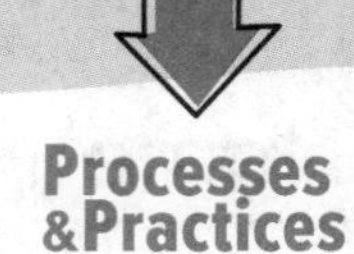

Use the data from the graph to answer the questions.

Favorite Pizza Toppings

Pepperoni	🍕	🍕	🍕	🍕	🍕
Sausage	🍕	🍕			
Cheese	🍕	🍕	🍕	🍕	
Vegetables	🍕	🍕	🍕		

Key: Each picture = 1 vote

10. Three more people were surveyed. They all like cheese. What topping is now the favorite?

____

Brain Builders

11. Carter voted for vegetables. Kylie voted for sausage. Whose favorite pizza topping got more votes? How many more votes?

____

Write Math Write a question about the graph above. Have a friend answer the question.

____

____

____

Name ....

My Homework

Lesson 3

Analyze Picture Graphs

Homework Helper

Need help? connectED.mcgraw-hill.com

You can use the data from a picture graph to answer questions.

Favorite Vacation

Beach					
Camping					
Water park					

Key: Each picture = 1 vote

Which vacation got the most votes? ____________

Practice

Use the data from the graph to answer the questions.

Favorite Hamburger Topping

Ketchup					
Mustard					
Lettuce					
Tomato					

Key: Each picture = 1 vote

1. How many people voted?

   ____

2. Which topping got the most votes? ____________________

3. Which two toppings got the least votes? __________________________

4. How many people like ketchup or tomato? ________

Use the data from the graph to answer the questions.

Favorite Sandwich

Peanut butter and jelly	🥪	🥪	🥪		
Turkey	🥪	🥪			
Grilled cheese	🥪	🥪	🥪		
Peanut butter	🥪	🥪	🥪	🥪	

Key: Each sandwich = 1 vote

Brain Builders

5. How many more people voted for turkey and grilled cheese together than voted for peanut butter and jelly? ________

6. Maria voted for peanut butter. She really wanted to vote for peanut butter and jelly. How would that change the picture graph?

____

____

____

7. **Test Practice** In the graph above, how many more people like grilled cheese than turkey?

1	2	4	7
○	○	○	○

Math at Home Create a picture graph for your child about your family's favorite activities. Ask your child questions about the data on the graph.

Name

Check My Progress

Vocabulary Check 

Write the word to complete each sentence.

picture graph **survey** **key**

1. In a ____________, you collect data by asking people the same question.

2. A ____________ tells what (or how many) each symbol stands for.

3. A ____________ has different pictures to show information collected.

Concept Check

Ask 10 people their favorite color. Use tally marks to record the data.

Use the data in the chart to answer the questions.

Favorite Color	Tally	Total
Blue		
Pink		
Green		
Purple		

4. What color do people like the most? ________

5. How many people like pink and green? ________

6. What color do people like the least? ________

Use the tally chart to make a picture graph.

7.

Favorite Toy	Tally
Jump rope	|||
Baseball	||||
Inline skates	||
Bike	~~||||~~

Favorite Toy

Jump rope					
Baseball					
Inline skates					
Bike					

Key: Each picture = 1 vote

Use the data from the graph to answer the questions.

8. How many more votes did bike get than baseball?

____

9. How many people like jump ropes and inline skates?

____

Brain Builders

10. How many more people voted for bike and jump rope all together than voted for baseball and inline skates all together?

____

11. If 3 more people vote for jump rope and 1 more person votes for baseball, which toy is the favorite?

____

12. **Test Practice** Two students did not vote. If those 2 students all voted for inline skates, which toy would have the most votes?

Jump rope ○ Baseball ○ Inline skates ○ Bike ○

Name ....

Lesson 4
Make Bar Graphs

ESSENTIAL QUESTION
How can I record and analyze data?

Math in My World

Favorite Winter Activity

Activity										
Sledding										
Ice Skating										
Skiing										
Snowboarding										
	0	1	2	3	4	5	6	7	8	9	10

Number

Our class likes ______________________ the best.

Teacher Directions: Ask 10 classmates to pick their favorite winter activity. Color one square for each vote in the correct row. Write your classmates' favorite winter activity.

Guided Practice

A **bar graph** uses bars to show data. To make a bar graph, color one box for each vote. Bar graphs can look different.

Favorite Breakfast Food	Tally	Total
Toast	||||	4
Cereal	~~||||~~	5
Eggs	||	2

One Way

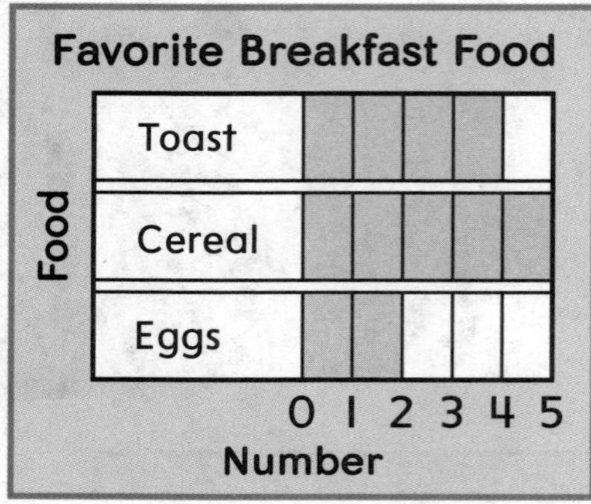

Another Way

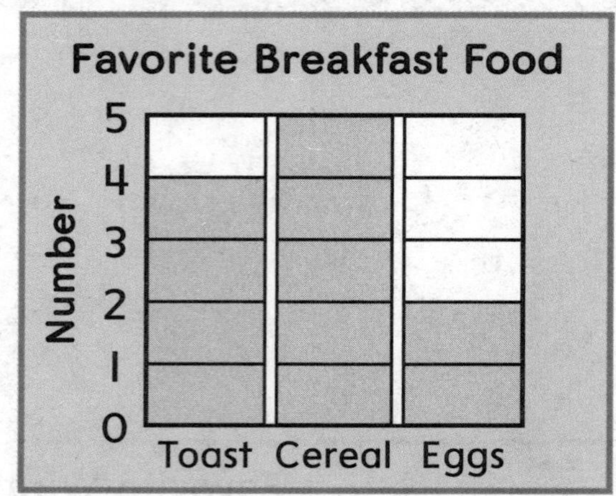

Use the tally chart to make a bar graph.

1.

Favorite Muffin	Tally	Total
Blueberry	||||	4
Strawberry	||	2
Lemon	|	1
Cherry	~~||||~~	5

Favorite Muffin

Muffin					
Blueberry					
Strawberry					
Lemon					
Cherry					
	0	1	2	3	4 5

Number

Talk Math How are bar graphs different from picture graphs?

Name ....

Helpful Hint
Color 1 box for each tally mark.

Independent Practice

Complete the tally chart. Use the tally chart to make a bar graph.

2.

Color of Shirt	Tally	Total
Red	||||	
Blue	𝍸 |	
Black	||	
Green	𝍸	

Color of Shirt

Colors						
Red						
Blue						
Black						
Green						

0 1 2 3 4 5 6

Number of Students

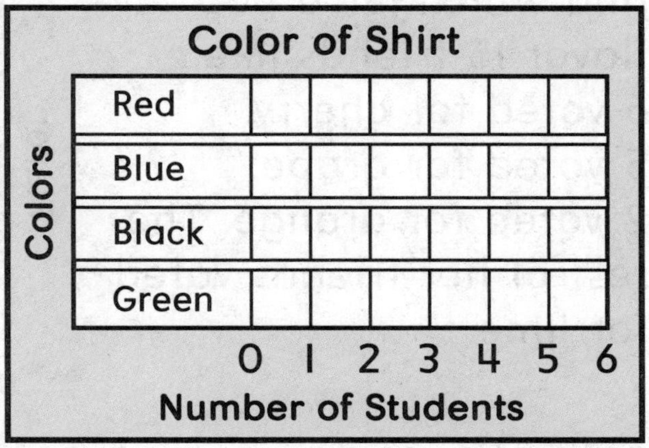

3.

Left-Handed or Right-Handed	Tally	Total
Left-Handed	||||	
Right-Handed	𝍸 |	

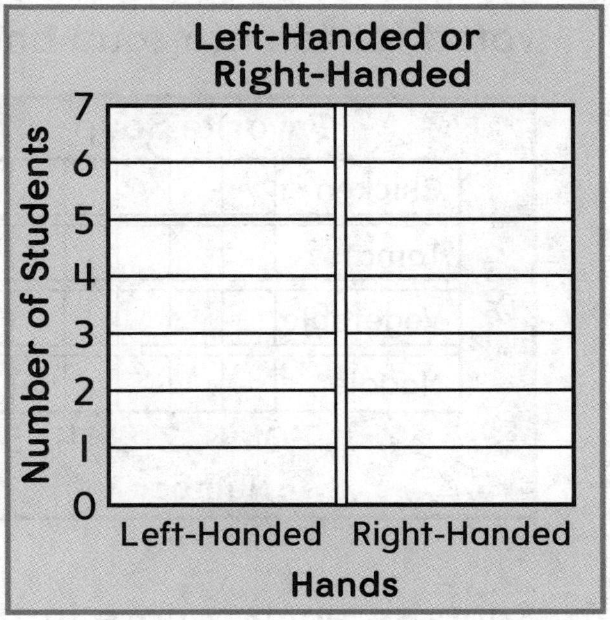

Left-Handed or Right-Handed

Number of Students	Left-Handed	Right-Handed
7		
6		
5		
4		
3		
2		
1		
0		

Hands

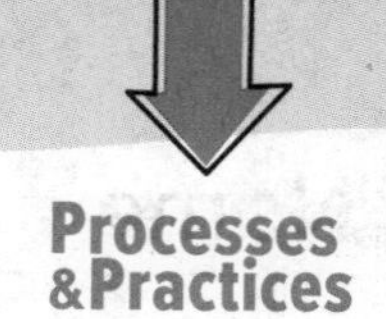

Processes & Practices

Brain Builders

Use the data to fill in the bar graphs.

4. Alyssa took a survey to find which juice pop flavor 15 friends like. 6 voted for cherry. 5 voted for grape. 2 voted for orange. The rest of her friends voted for lime.

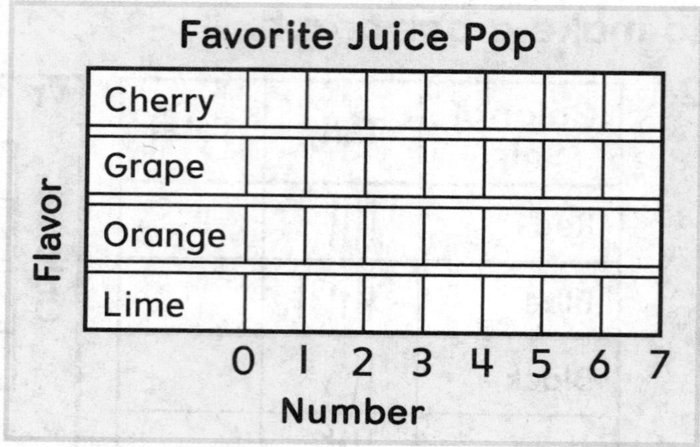

5. Luke took a survey of 18 friends to find their favorite soup. 4 voted for chicken soup. 6 voted for vegetable. The same number voted for tomato soup and noodle soup.

6. Suppose each square in the Favorite Soup graph above was worth 2 votes. How would your bar graph change?

____

____

Name

My Homework

Lesson 4

Make Bar Graphs

Homework Helper

Need help? connectED.mcgraw-hill.com

You can use data from a tally chart to make a bar graph.

Favorite Music	Tally	Total
Country	|||	3
Rock	||||	4
Jazz	||	2

Practice

Helpful Hint
Color 1 box for each tally mark.

Use the tally chart to make a bar graph.

1.

Favorite Bird	Tally	Total
Robin	|||	3
Blue Jay	||	2
Swan	||||	4
Flamingo	~~||||~~	5

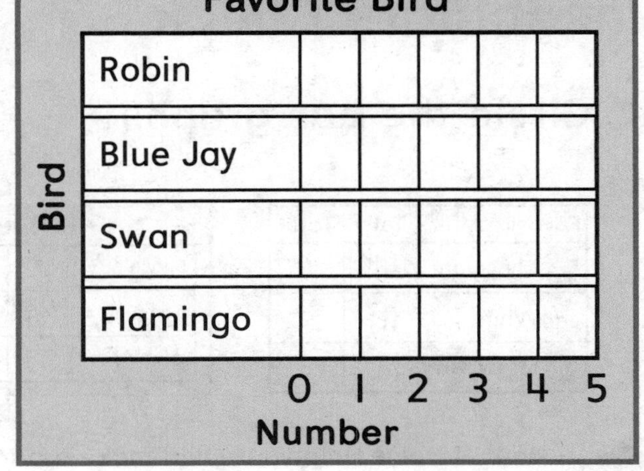

Use the tally chart to make a bar graph.

2.

Favorite Color	Tally	Total
Red	||||	4
Blue	~~||||~~	5
Pink	||||	4
Green	||	2

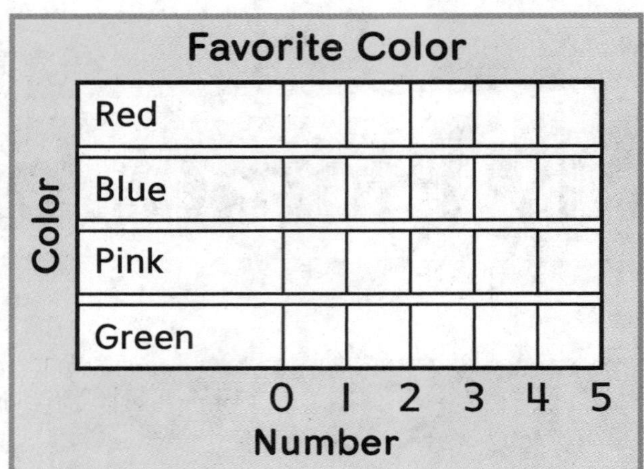

Brain Builders

Use the data to fill in the bar graph.

3. 15 people voted on their favorite field trip. 2 voted for the symphony. 5 voted for the zoo. The same number voted for the museum and the aquarium.

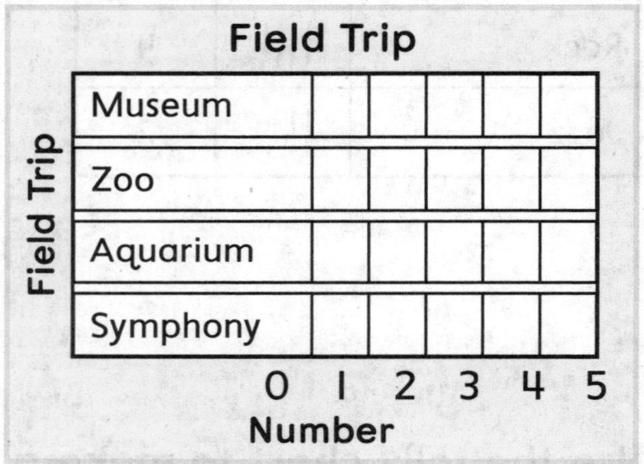

Vocabulary Check

4. Circle the **bar graph.**

Favorite Milk	Tally	Total
Chocolate	~~||||~~	5
White	||	2
Strawberry	|||	3

Favorite Vacation

Beach					
Camping					
Water park					

Key: Each picture= 1 vote

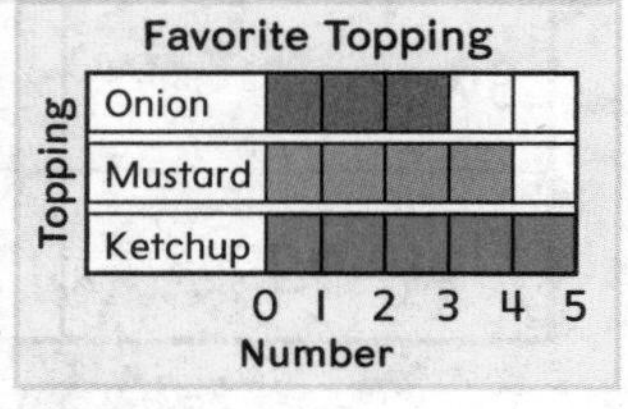

Math at Home Help your child make a bar graph of the type of weather he or she sees for a week.

Name

Lesson 5

Analyze Bar Graphs

ESSENTIAL QUESTION

How can I record and analyze data?

Math in My World

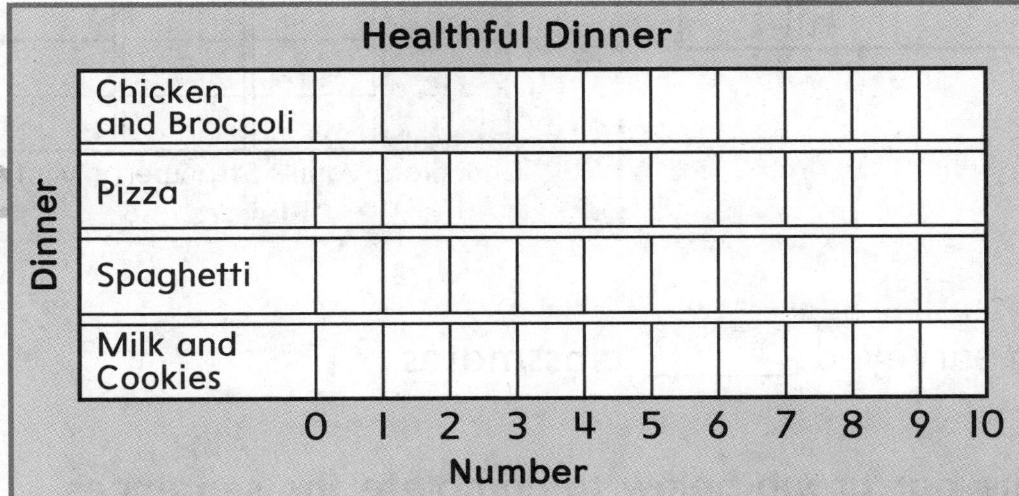

chicken and broccoli ________

pizza ________

spaghetti ________

milk and cookies ________

Teacher Directions: Survey 10 people. Ask them which dinner is the healthiest. Make a bar graph. Write how many votes each dinner got.

Guided Practice

Owen surveyed his class about their favorite ice cream flavors. Then he made a bar graph to show the data.

Favorite Ice Cream Flavor	Tally
Chocolate	𝍸 ||
Vanilla	|||
Strawberry	𝍸
Mint	𝍸 |

Owen surveyed 21 classmates.

Use the bar graph below to complete the sentences.

1. The flavor with the least votes is ___________.

2. Chocolate received 2 more votes than___________.

3. What two flavors did a total of 11 students vote for?

____

Talk Math How could this graph help a teacher plan a class party?

Name

Independent Practice

Use the bar graph to answer the questions.

4. What color hair do 5 students have?

____

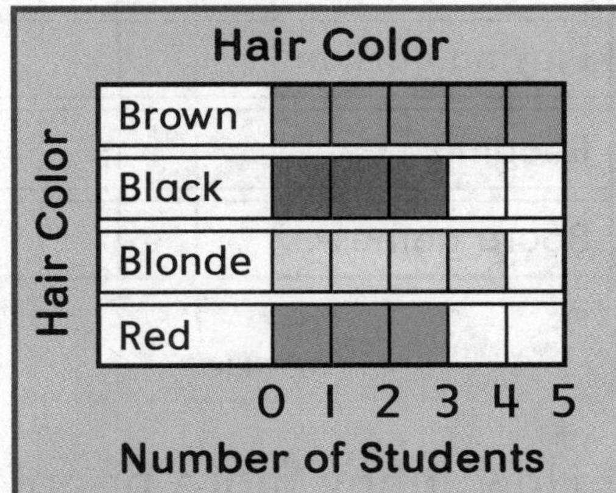

5. What color hair do 4 students have?

____

6. What two hair colors do the same number of students have?

____

7. How many students were surveyed? __________

8. How would this graph change if 20 students took the survey?

____

9. Which pet do most students have?

____

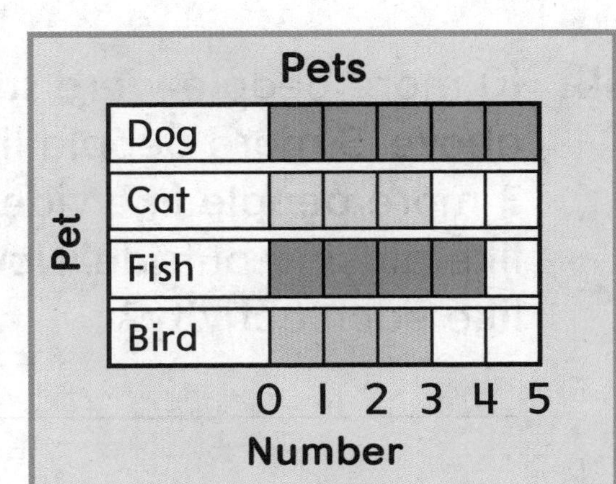

10. Which two pets do the fewest students have?

____

11. How many students have cats and birds? __________

Problem Solving

Use the bar graph to answer the questions.

Favorite Activity

Activity					
Video games					
Playing outside					
Reading					
Board games					
	0	1	2	3	4 5

Number

12. How many more people like reading or board games than playing video games?

______

Brain Builders

13. Samantha also chooses an activity. Then 3 activities are the favorite of the same number of people. Which activity did she choose?

____

14. 10 more people were surveyed for the graph above. 3 more people like board games. 3 more people like video games. The rest like playing outside. Now how many people like each activity?

____

____

____

Name ....

My Homework

Lesson 5

Analyze Bar Graphs

Homework Helper

Need help? connectED.mcgraw-hill.com

You can answer questions using the data from a bar graph.

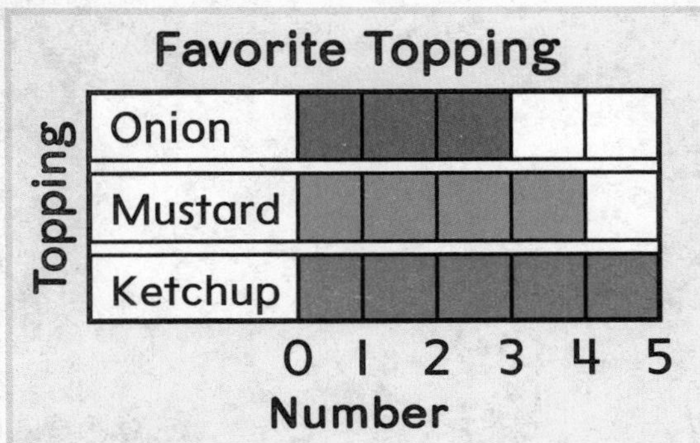

How many students were surveyed?

12 students were surveyed.

Practice

Use the bar graph to answer the questions.

1. How many people voted for mustard?

   ____

2. What topping do most people like?

   ____

3. Which topping is liked the least?

   ____

4. How many people like mustard or onion? ________

Use the bar graph to answer the questions.

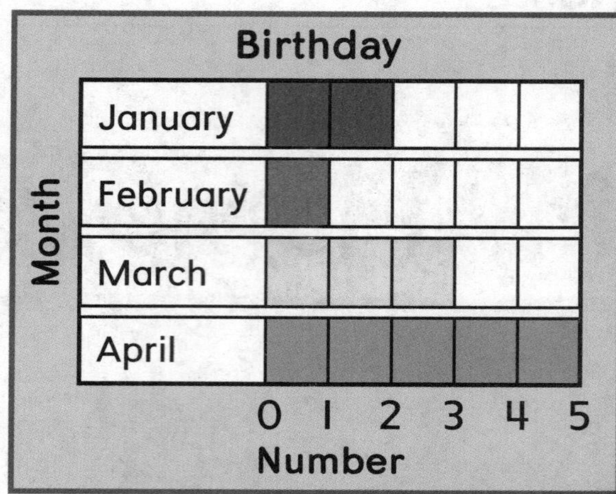

5. How many people have April birthdays?

______

6. How many birthdays are in February and March?

______

7. How many people were surveyed in all?

______

Brain Builders

8. 4 more people answered the survey. They all have birthdays in the same month. Now that month has 5 people. What month are their birthdays in?

______

9. **Test Practice** How many birthdays are in May and August?

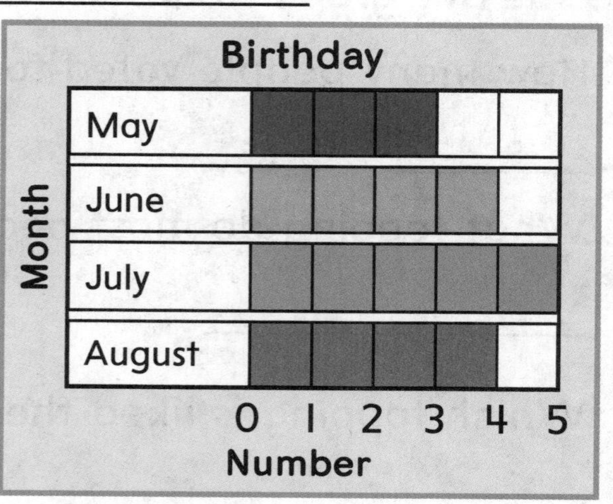

8 ○ 7 ○

4 ○ 3 ○

Math at Home Ask your child how he or she analyzed the bar graphs on this page.

Name

Lesson 6

Problem Solving

STRATEGY: Make a Table

ESSENTIAL QUESTION
How can I record and analyze data?

Watch

The breakfast special at Darla's Diner comes with 3 pancakes. Five friends order the breakfast special. How many pancakes will they get in all?

1 Understand Underline what you know. Circle what you need to find.

2 Plan How will I solve the problem?

3 Solve Make a table.

Friends	Pancakes
1	3
2	6
3	9
4	12
5	15

15 pancakes

4 Check Is my answer reasonable? Explain.

Practice the Strategy

Each person in the class has the same number of pets. Two people have 4 pets. How many pets do 7 people have?

1 Understand Underline what you know. Circle what you need to find.

2 Plan How will I solve the problem?

3 Solve I will...

Number of Pets	
People	Pets
1	2
2	4
3	
4	
5	
6	
7	

_____ pets

4 Check Is my answer reasonable? Explain.

Name ....

Processes & Practices

Apply the Strategy

1. Desiree has 4 pairs of socks in her drawer. How many socks are there in all?

Pair	Socks in All
1	2

______ socks

Brain Builders

2. Mr. Minnick needs to deliver 60 boxes. His car can hold 10 boxes at a time. How many trips will he need to make to deliver all 60 boxes? Explain.

____

____

____

3. Juice boxes come in packs of 4 at Pat's Grocery. Mrs. Perez needs 22 juice boxes in all. How many packs should she buy?

______ packs

Review the Strategies

Choose a strategy
- **Make a table.**
- **Make a model.**
- **Find a pattern.**

4. Aaron has 3 muffin pans. Each pan can hold 6 muffins. How many muffins can Aaron bake at a time?

_____ muffins

5. Suzie, Simon, and Sean each had 4 bottles of water during their softball game. How many bottles of water did they have in all?

_____ bottles

6. Mr. Bell wants his 21 students to put on their gloves to go outside. How many gloves do they have in all?

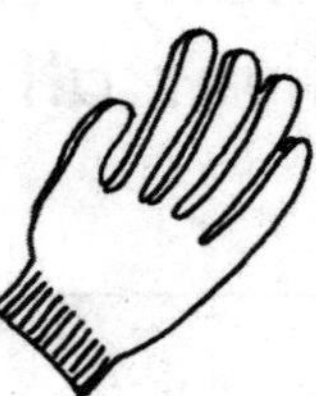

_____ gloves

7. Grace, Elijah, and Cameron are saving soup labels. Grace has 10, Elijah has 7, and Cameron has 13. How many more labels does Cameron have than Elijah?

_____ labels

Name

My Homework

Lesson 6

Problem Solving: Make a Table

Homework Helper

Need help? connectED.mcgraw-hill.com

Each hen lays the same number of eggs. One hen lays 3 eggs. Two hens lay 6 eggs. Three hens lay 9 eggs. How many eggs will 6 hens lay?

1 Understand Underline what you know. Circle what you need to find.

2 Plan How will I solve the problem?

3 Solve I will make a table.

Each hen lays 3 eggs. So, 6 hens will lay 18 eggs.

Hens	Eggs
1	3
2	6
3	9
4	12
5	15
6	18

4 Check Is my answer reasonable?

Problem Solving

Underline what you know. Circle what you need to find.

1. Four turtles fit in one tank. Juan has 3 tanks. How many turtles can he have?

_____ turtles

2. Nine children want to feed the birds. They each have 2 bags of seed. How many bags of seed are there in all?

_____ bags

Brain Builders

3. Shandra is giving a fruit snack bag to each of her 4 friends. She puts 4 pear slices and 2 apple slices in each bag. How many fruit slices are there in all?

_____ fruit slices

4. 8 children and 2 adults are each making a snowman. Each snowman needs 3 snowballs. How many snowballs do they need in all?

_____ snowballs

Math at Home Have your child make a table to show how many meals he or she eats in a week.

Name

Lesson 7
Make Line Plots

ESSENTIAL QUESTION
How can I record and analyze data?

Math in My World

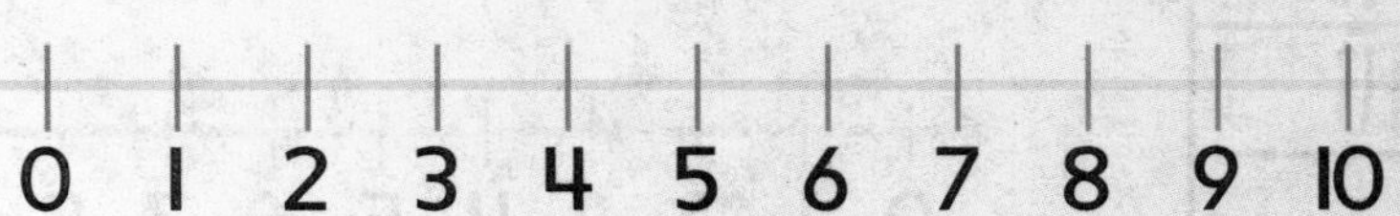

Teacher Directions: Chase asked 10 students how many servings of fruits and vegetables they eat each day. Three people eat 2 servings. Draw 3 X's over the number 2. Six people eat 5 servings. Draw 6 X's over the number 5. One person eats 7 servings. Draw 1 X over the number 7.

Guided Practice

A **line plot** is a way to organize data. Line plots are used to see how often a certain number occurs in data.

Mr. Sun's students marked their ages on a tally chart. They made a line plot using the data.

Age	Tally
7	|||
8	𝍸
9	||

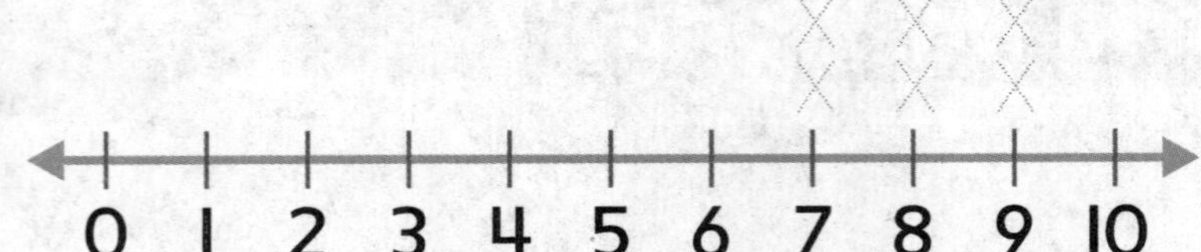

Use the tally chart to make a line plot.

1.

Pets	Tally
1	||
2	|||
3	||
4	|

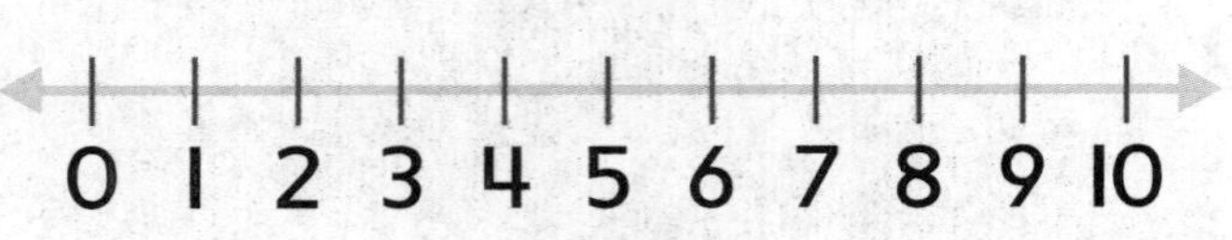

2.

Sisters	Tally
0	||
1	|||
2	
3	|

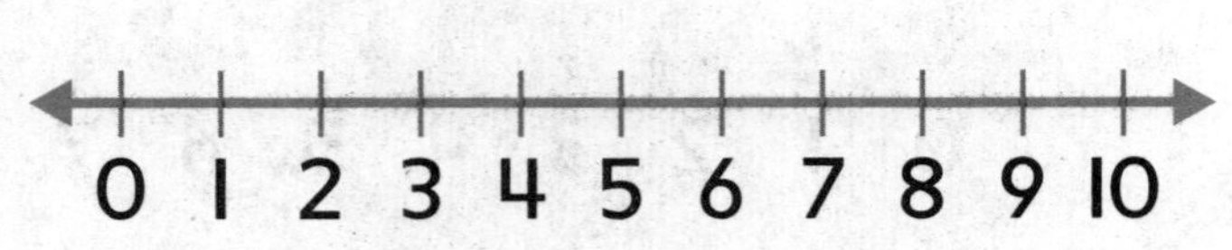

Talk Math How are line plots similar to tally charts?

Name ....

Independent Practice

Use the tally chart to make a line plot.

3.

Brothers	Tally
0	|||
1	||||
2	
3	|||
4	|

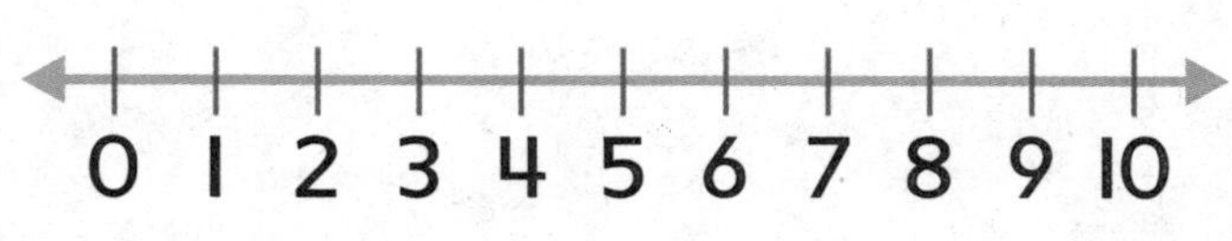

4.

Cousins	Tally
4	|||
5	||||
6	卌 |
7	卌

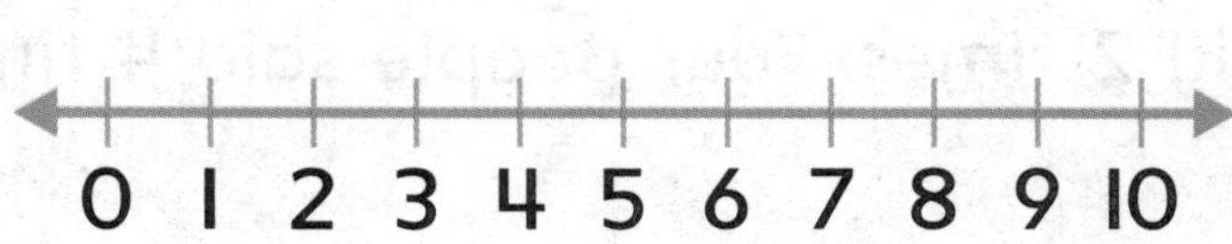

5.

Aunts	Tally
3	卌 |
4	|||
5	||
6	|||

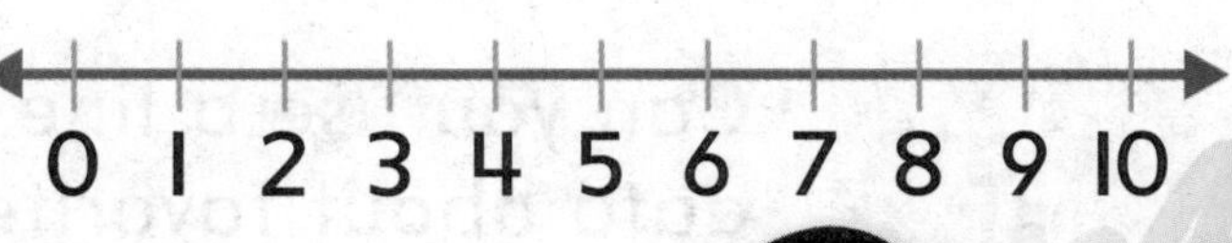

Processes & Practices

Brain Builders

Use the data to make a line plot.

6. Eli asked 10 classmates how many glasses of water they drink each day. Two people said 1. Three people said 0. Three people said 4. Everyone else said 6.

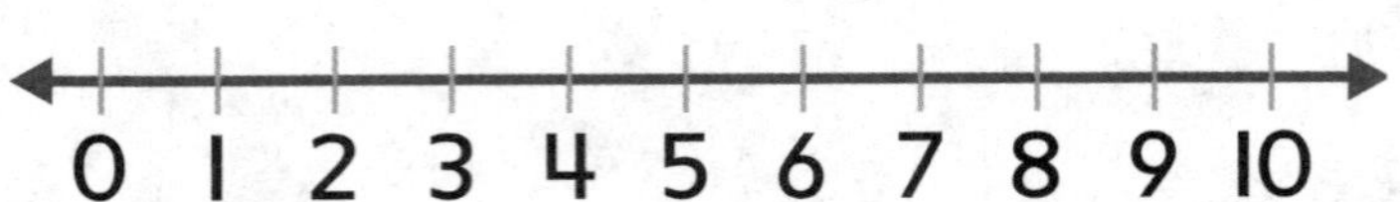

7. Sydney asked 15 friends how many times they exercise each week. The same number of people said 1 time and 3 times. Five people said 2 times. Four people said 4 times.

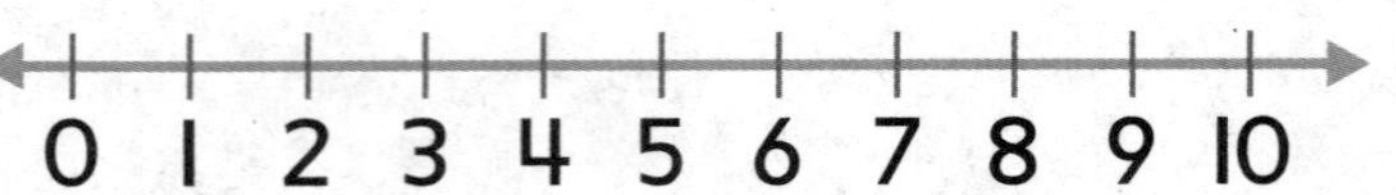

Write Math Can you use a line plot to show data about favorite color?

____

____

____

Name ....

Lesson 7

Make Line Plots

My Homework

Homework Helper

Need help? connectED.mcgraw-hill.com

You can use data from a tally chart to make a line plot. Line plots show how often a certain number occurs in data.

Sports	Tally
0	||
1	|||
2	𝍸
3	|||

0 1 2 3 4 5

Practice

Use the tally chart to make a line plot.

1.

Siblings	Tally
0	|||
1	|
2	|||
3	||

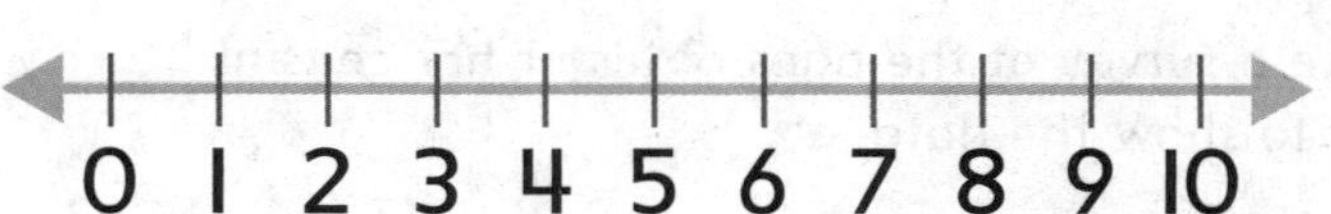

Use the tally chart to make a line plot.

2.

Second Grade Classes	Tally
1	𝍷
2	𝍷𝍷𝍷
3	𝍸
4	𝍸𝍷𝍷

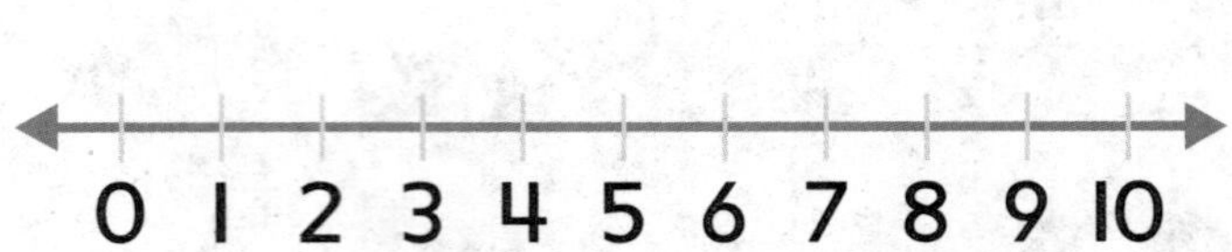

Brain Builders

3.

Swings	Tally
5	𝍷𝍷𝍷𝍷
6	𝍸𝍷𝍷
7	𝍷𝍷𝍷
8	𝍸𝍷

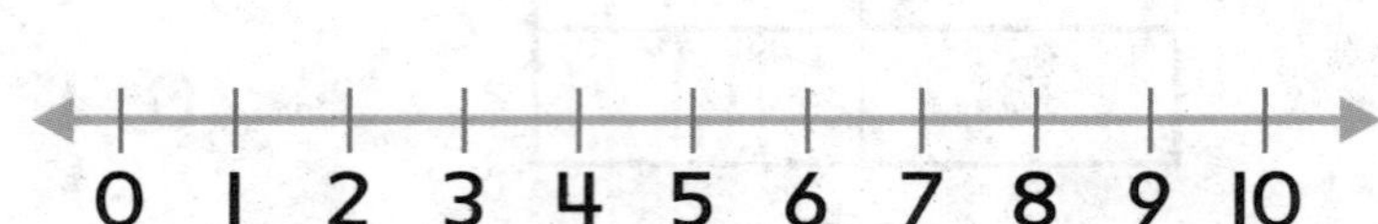

Explain to a friend or family member how you made the line plot.

Vocabulary Check

4. Circle the **line plot**.

Our Favorite Toys

Balls

Skates

Stuffed animals

Math at Home Help your child take a survey of the ages of his or her cousins. Have your child make a line plot to show the data.

Name ________________

Lesson 8
Analyze Line Plots

ESSENTIAL QUESTION
How can I record and analyze data?

Math in My World

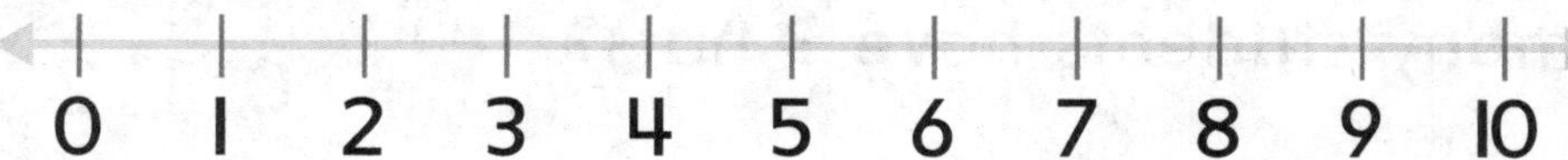

How many times a week do most people eat dessert? ________________

Teacher Directions: Ask 10 people how many times they eat dessert each week. Use the data to make a line plot. Answer the question.

Processes &Practices

Guided Practice

You can use data from a line plot to answer questions.

This line plot shows the number of students in each class at a school.

Number of Students

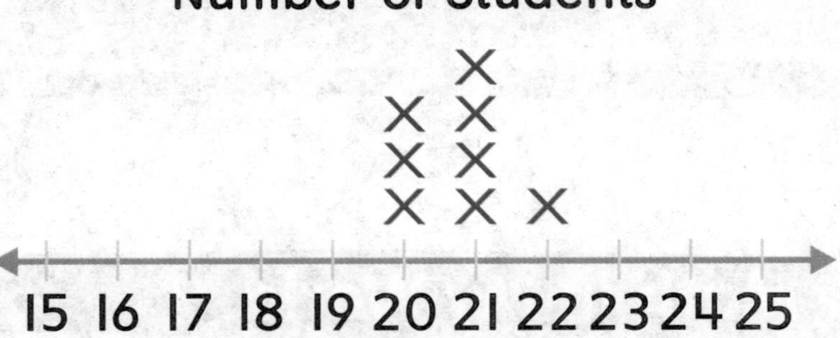

How many students are in most of the classes? 21

Use the data from the line plot to answer the questions.

Number of Hats

0 1 2 3 4 5

1. How many students have 2 hats?

_____

2. How many students have just 1 hat?

_____

3. How many students have 3 hats?

_____

4. How many students have more than 2 hats?

_____

Talk Math How are line plots the same or different than bar graphs or picture graphs?

Name

Independent Practice

Use the data from the line plot to answer the questions.

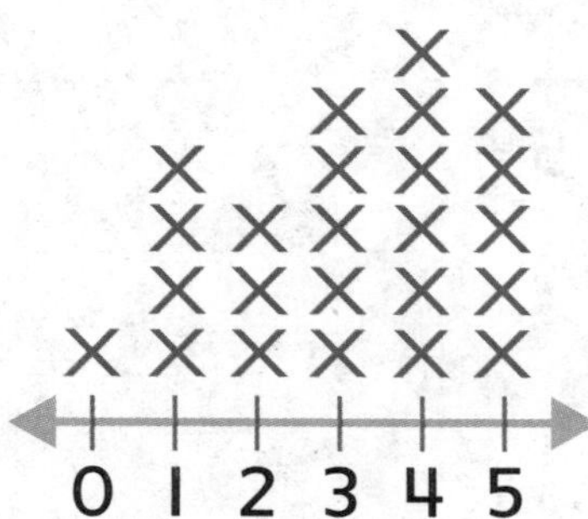

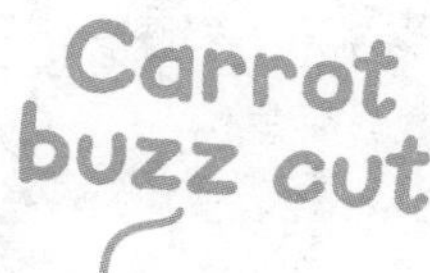

5. Which number has the most Xs? ______

6. How many students like 4 vegetables? ______

7. Which number has the fewest Xs? ______

8. How many students like 2 vegetables? ______

9. How many students were surveyed in all? ______

10. How many students like more than 1 vegetable?
______ students

11. How many students do not like vegetables?
______ student(s)

Problem Solving

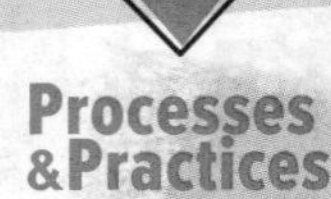

12. Finish the line plot for Ashley. She needs to show that three friends each have 2 toy bears.

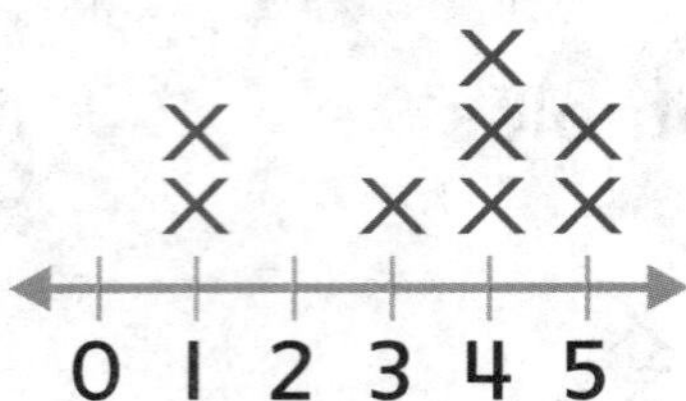

Brain Builders

13. 23 students created this line plot to show how many fruits each student likes. 6 students like the same number of fruit. How many fruits does each of those 6 students like?

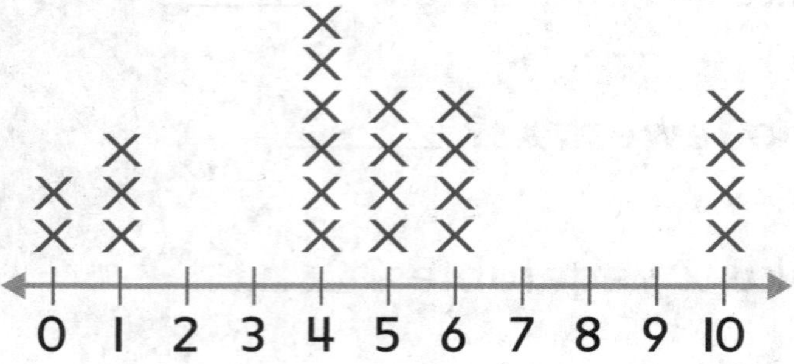

______ fruits

Write Math Colton looks at the line plot above and says 1 student likes three fruits. Tell why he is wrong. Make it right.

____

____

____

Name ....

My Homework

Lesson 8

Analyze Line Plots

Homework Helper

Need help? connectED.mcgraw-hill.com

You can answer questions from data in line plots.

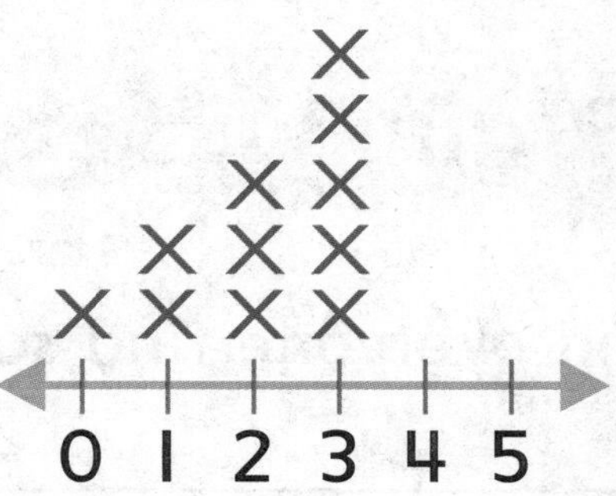

How many friends have 3 dolls?

5 friends each have 3 dolls.

Practice

Use the data from the line plot to answer the questions.

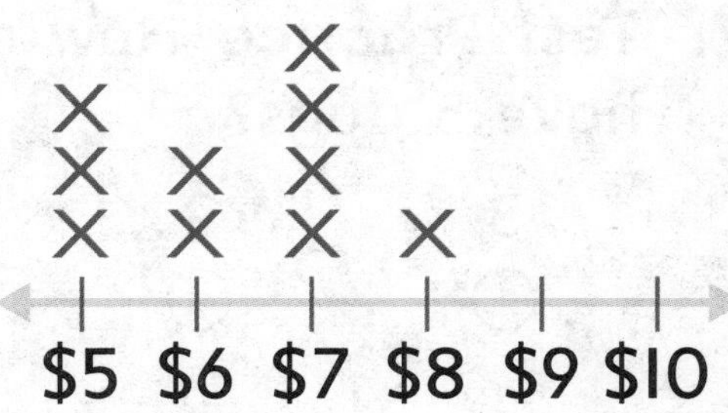

1. How many people get $5? ______

2. How much allowance do most people get?

______

3. How many people get more than $6?

______

Use the data from the line plot to answer the questions.

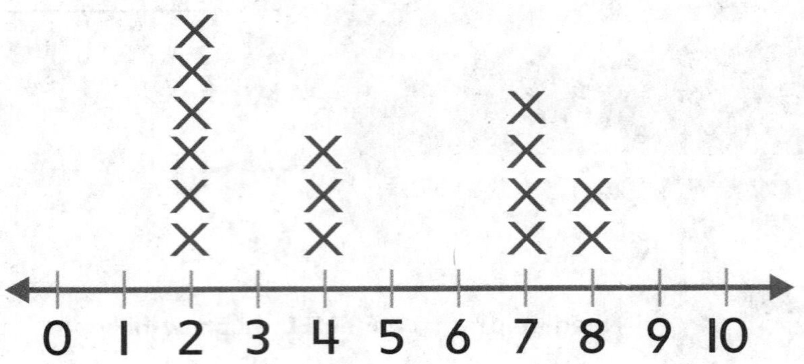

4. How many people found pennies? ______

Brain Builders

5. Did everyone find some pennies?

______

______

6. How many people found less than 5 pennies? Explain to a friend or family member how you know.

______

7. **Test Practice** How many people have 5 dogs?

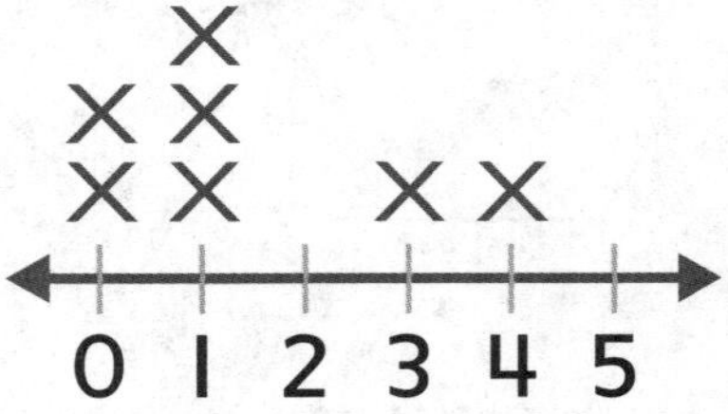

○ 3
○ 2
○ 1
○ 0

Math at Home Make a line plot about the number of healthy food choices your child makes each day. Ask your child questions about the line plot at the end of a week.

Name ....

My Review

Chapter 9

Data Analysis

Vocabulary Check

data	survey	tally mark
picture graph	line plot	bar graph

Write the correct word in each blank.

1. You can collect data by asking people the same question in a ____________.

2. A ____________ uses bars to show data.

3. Numbers or symbols that show information are called ____________.

4. A ____________ shows how often a certain number occurs in data.

5. A ____________ uses pictures to show information collected.

6. A ____________ is a mark used to record data collected in a survey.

Concept Check

Use the tally chart to make a picture graph.

7.

<table>
<tr><th colspan="2">Favorite Sport</th></tr>
<tr><th>Sport</th><th>Tally</th></tr>
<tr><td>Baseball</td><td>||||</td></tr>
<tr><td>Football</td><td>||||</td></tr>
<tr><td>Hockey</td><td>||</td></tr>
<tr><td>Volleyball</td><td>|||</td></tr>
</table>

Favorite Sport

Baseball					
Football					
Hockey					
Volleyball					

Key: Each picture = 1 vote

Use the picture graph above to complete the sentences.

8. How many students were surveyed? ________

9. How many students like hockey or football? ________

Use the tally chart to make a bar graph.

10.

<table>
<tr><th>Have you been to the zoo?</th><th>Tally</th></tr>
<tr><td>Yes</td><td>卌 卌</td></tr>
<tr><td>No</td><td>|</td></tr>
</table>

Have you been to the zoo?

Yes										
No										

0 1 2 3 4 5 6 7 8 9 10

Number of Students

Use the bar graph above to complete the sentences.

11. How many students have been to the zoo? ________

12. How many more students have been to the zoo than have not been to the zoo? ________

Name ....

Problem Solving

13. Taylor asked 12 friends to name their favorite number. Five friends said 4. Six friends said 1. One friend said 2. Use Taylor's data to complete the line plot.

Our Favorite Numbers

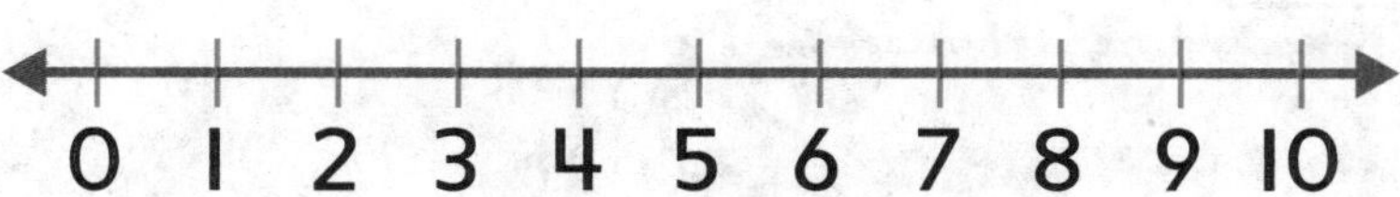

14. A tricycle has 3 wheels. A bike has 2 wheels. How many wheels are there on 4 tricycles and 3 bikes?

______ wheels

15. **Test Practice** There are 4 giraffes at the zoo. Each giraffe has 4 legs. How many legs are there in all?

2	4	8	16
○	○	○	○

Reflect

Chapter 9

Answering the Essential Question

Show how to record and analyze data.

Complete the tally chart.

Fruit	Tally	Total
Banana	卌	
Apple	|||||	
Orange	||||	

Use the tally chart to make a picture graph.

Fruit

Banana					
Apple					
Orange					

Key: Each picture = 1 vote

ESSENTIAL QUESTION

How can I record and analyze data?

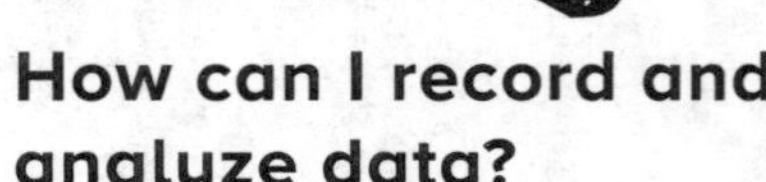

Use the tally chart to make a bar graph.

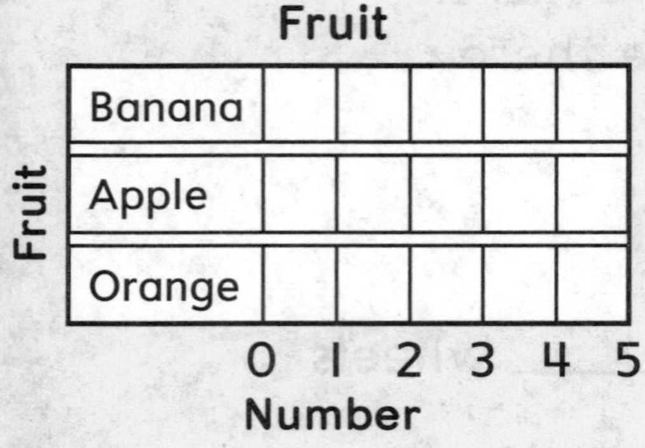

Analyze the data.

How many people like apples or bananas?

______

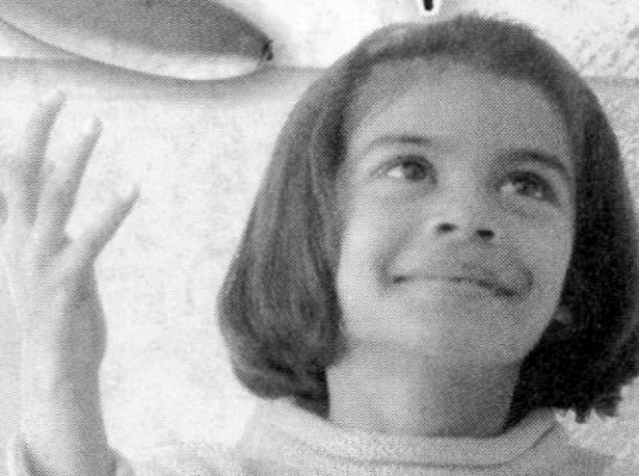

Keep your eyes on your goal!

Name .... Date ....

Score ....

Performance Task

Letter Counts

Research how many letters are in the names of the months in a calendar year. Record your research in the following table, using tally marks.

Number of Letters	Tally

Show all your work to receive full credit.

Part A

Complete the following bar graph.

Letters in Months' Names

Number: 0, 1, 2, 3, 4, 5, 6, 7

5 or fewer | 6 | 7 or more

Part B

Using the bar graph in **Part A**, can you tell that 8 is the most common number of letters in a month's name? Explain.

Part C

Create a line plot from the data you collected in **Part A**. Be sure to use labels.

Chapter

10 Time

ESSENTIAL QUESTION

How do I use and tell time?

Away We Go!

Watch a video!

Name

Chapter 10 Project

Create a Temperature Log

1. Write the date and time on your Temperature Log. Then record the temperature shown on the thermometer.
2. At the end of the chapter, write a paragraph comparing the warmest day and the coldest day below.
3. Subtract to find the difference in temperature between the warmest and coldest days. Record the difference.

Compare the warmest day you recorded to the coldest day you recorded below.

The difference in the temperature that I recorded is ______.

Name ....

Am I Ready?

Circle the time.

1. 

7 o'clock 2 o'clock 6 o'clock

2. 

4:00 5:00 6:00

Find the missing numbers.

3. 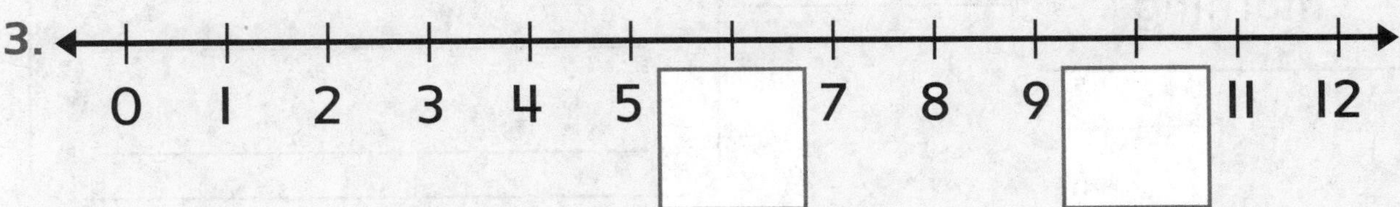

4. 2, 4, ______, 8, ______, 12, 14, 16, ______, 20

5. 5, ______, ______, 20, ______, ______, 35, 40, ______

6. Kira is walking home from school. Circle the time of day for the activity.

morning afternoon evening

How Did I Do?

Shade the boxes to show the problems you answered correctly.

1	2	3	4	5	6

Name

My Math Words

Review Vocabulary

afternoon evening morning

Write about an activity you do for each time of day. Use *morning, afternoon,* or *evening* in each sentence.

morning

afternoon

evening

My Vocabulary Cards

Processes & Practices

Lesson 10-6

A.M.

Lesson 10-1

analog clock

Lesson 10-1

digital clock

Lesson 10-2

half hour

half past 5
5:30

Lesson 10-1

hour

3:00

hour

3 o'clock
3:00

Lesson 10-1

hour hand

Teacher Directions:
Ideas for Use
Directions:

- Tell students to create riddles for each word. Ask them to work with a friend to guess the word for each riddle.
- Ask students to arrange cards to show an opposite pair. Have them explain the meaning of their pairing.

A clock that has an hour hand and a minute hand.

The hours from midnight until noon.

One half hour is 30 minutes. Sometimes called half past.

A clock that uses only numbers to show time.

The shorter hand on a clock that tells the hour.

A unit to measure time. 1 hour = 60 minutes

My Vocabulary Cards

Lesson 10-2

Lesson 10-1

Lesson 10-6

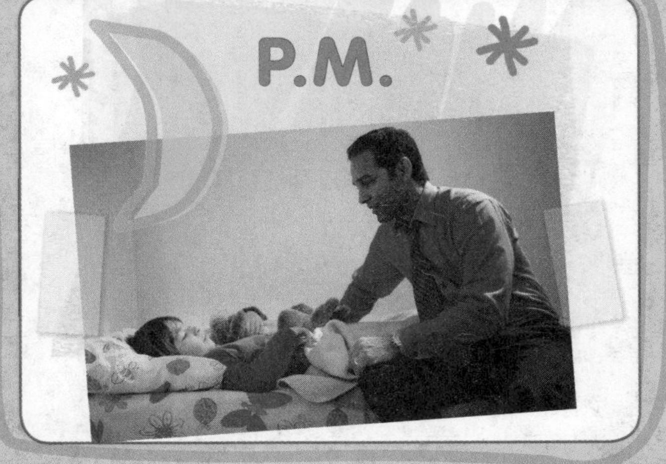

Lesson 10-4

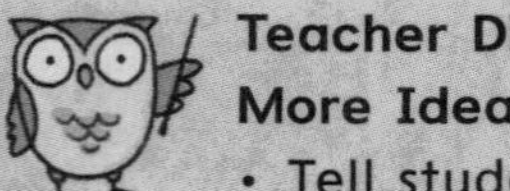

Teacher Directions:
More Ideas for Use

- Tell students to draw a picture to show the meaning of a word. Have them work with a friend to guess the word.
- Have students use a blank card to write this chapter's essential question. On the back, have them write or draw examples that help them answer the question.

The longer hand on a clock that tells the number of minutes.

A unit to measure time. Each tick mark on an analog clock.
1 minute = 60 seconds
60 minutes = 1 hour

A quarter hour is 15 minutes. Sometimes called quarter past or quarter till.
4 quarter hours = 1 hour

The hours from noon until midnight.

My Foldable

FOLDABLES® Follow the steps on the back to make your Foldable.

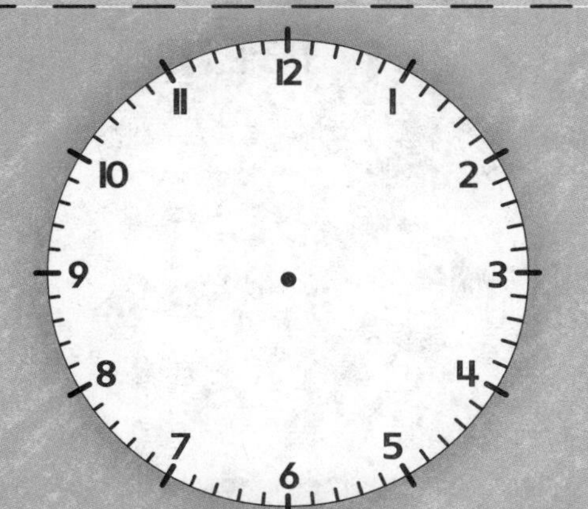

A.M. P.M.

A.M. P.M.

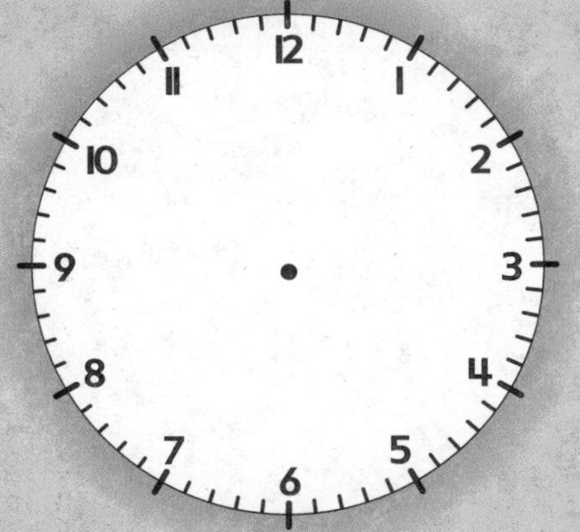

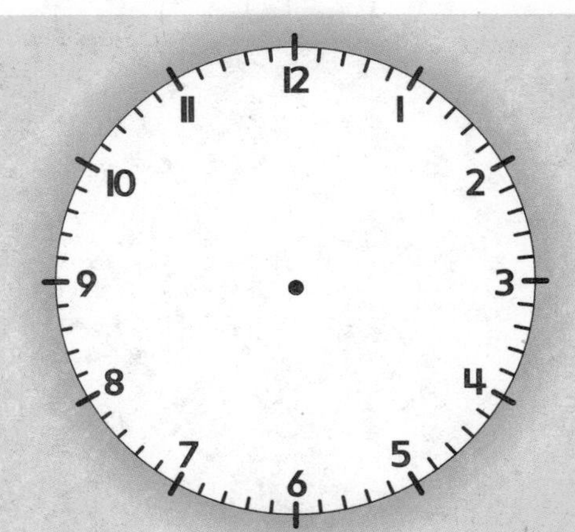

A.M. P.M.

A.M. P.M.

1

2

3

1 o'clock

quarter till 9

quarter past 2

half past 3

Name

Lesson 1
Time to the Hour

ESSENTIAL QUESTION
How do I use and tell time?

Math in My World

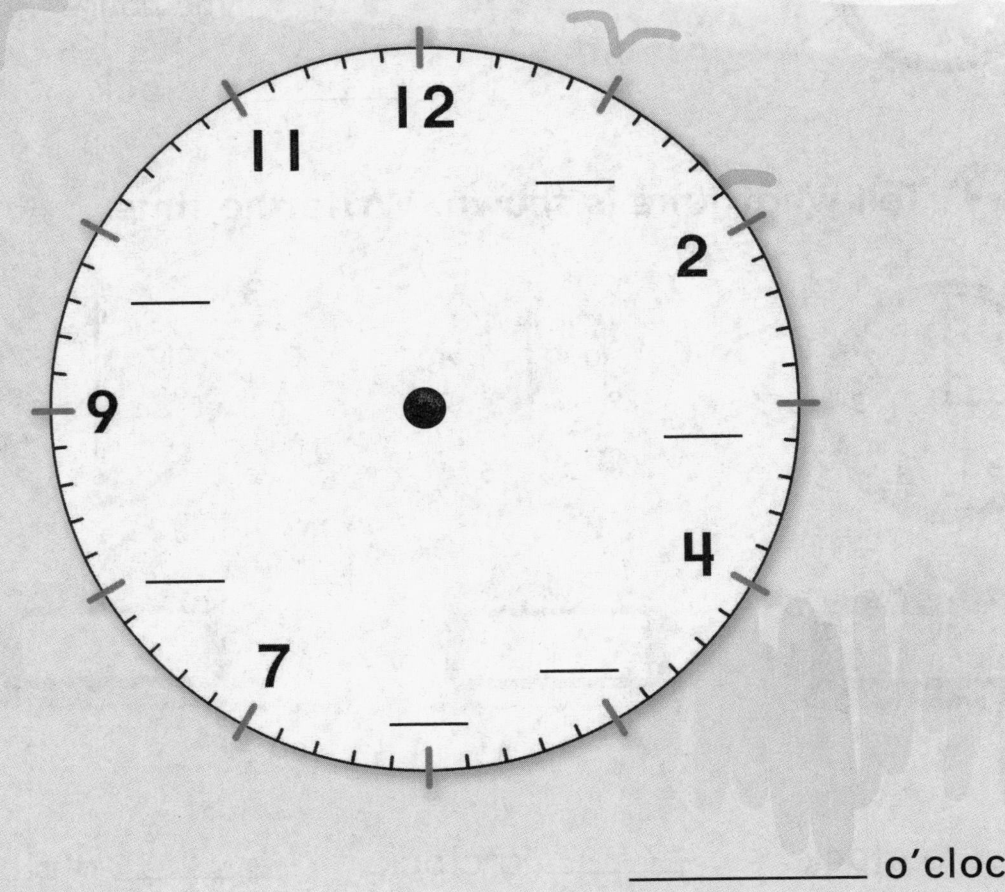

_______ o'clock

Teacher Directions: Write the missing numbers on the clock. Use [clock] to show time to each hour. Draw hands on the clock to show 4 o'clock. Write the time.

Processes & Practices

Guided Practice

On an **analog clock**, the **hour hand** points to the **hour**. It is shorter.

The **minute hand** points to the **minute**. It is longer.

A **digital clock** shows the hour and minutes on a screen.

3 o'clock

Use [clock]. Tell what time is shown. Write the time.

1.

_____ o'clock

2.

_____ o'clock

3.

_____ o'clock

Talk Math How is an analog clock similar to a number line?

Name ....

Independent Practice

Use [clock]. Tell what time is shown. Write the time.

4. 

_______ o'clock

5. 

_______ o'clock

6. 

_______ o'clock

Draw the hands on the clock. Write the time.

7. 5 o'clock

8. 10 o'clock

9. 3 o'clock

Problem Solving

Processes & Practices

Show the time on each clock.

10. Colin gets home at 3 o'clock. Evan gets home one hour later. What time does Evan get home?

11. Basketball practice starts at 4:00. Mark eats a snack 2 hours before practice. What time does Mark eat his snack?

Brain Builders

12. It is 4 hours before 12 o'clock. What time is it if it is exactly on the hour? Show the time on the clock.

Write Math What time is it when the hour hand is on 5 and the minute hand is on 12? How do you know?

____

____

____

Name

Lesson 1

Time to the Hour

My Homework

Homework Helper

Need help? connectED.mcgraw-hill.com

You can tell and write time to the hour.

It is 5 o'clock or 5:00.

Practice

Tell what time is shown. Write the time.

1.

_______ o'clock

2.

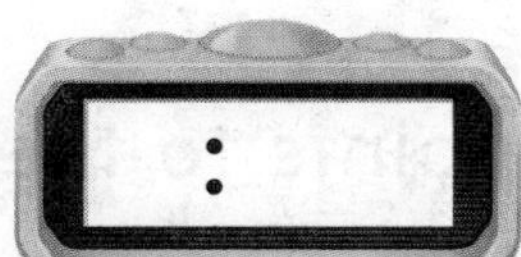

_______ o'clock

3.

_______ o'clock

Draw the hands on each clock. Then write the time.

4. 2 o'clock

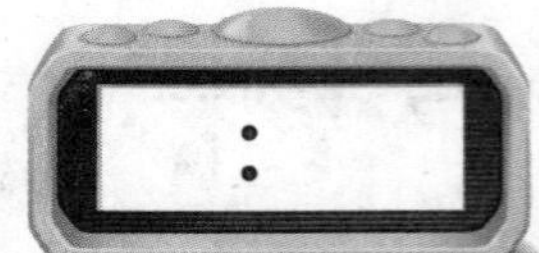

5. 4:00

Brain Builders

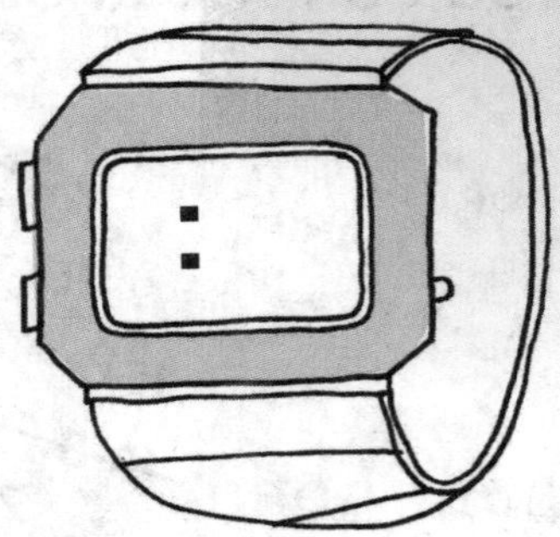

6. The time on Javier's watch now is 3 hours after 2:00. Write the time on Javier's watch that it will show in 1 hour.

Vocabulary Check

Complete each sentence.

hour hand **analog clock** **minute hand** **digital clock**

7. An ______________ has an hour and a minute hand.

8. The ______________ points to the hour on an analog clock.

9. A ______________ uses numbers on a screen to show time.

10. The ______________ points to the minute on an analog clock.

Math at Home Throughout the day, ask your child to look at an analog or digital clock showing a time to the hour and have them tell you the time.

Name

Lesson 2
Time to the Half Hour

ESSENTIAL QUESTION
How do I use and tell time?

Math in My World

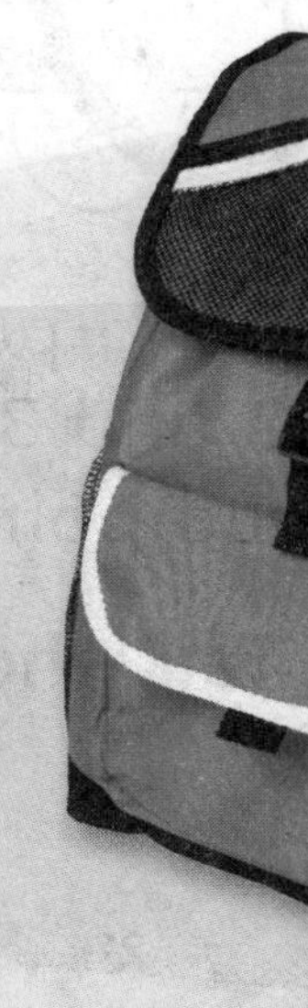

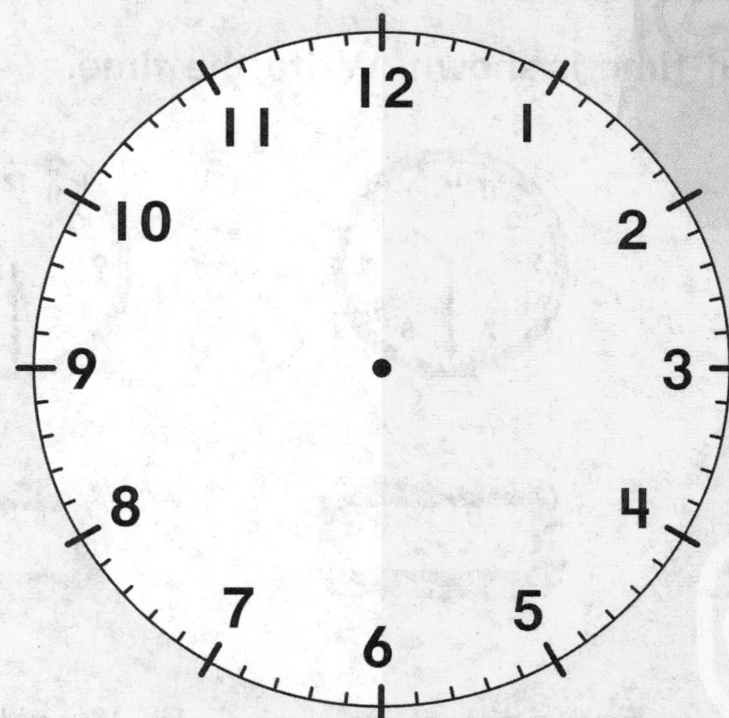

half past ________

Teacher Directions: Use a [clock] to show 5 o'clock. Move the minute hand to 6. What time is it? Repeat with other times showing hour and half past the hour. Draw hands to show one of the times and write the time.

Processes &Practices

Guided Practice

You can show time to the half hour.
One half hour is 30 minutes.
These clocks show **half past** 2.

> **Helpful Hint**
> The hour hand moves halfway to the next number as a half hour passes.

The hour hand points between 2 and 3. The minute hand points to 6.

It is 2:30.

Use  . Tell what time is shown. Write the time.

1. 

half past ______

2. 

half past ______

3. 

half past ______

Talk Math It is half past 8. Explain what *half past* means.

Name

Independent Practice

Use . Tell what time is shown. Write the time.

4.

half past ______

5.

half past ______

6.

half past ______

Draw the hands on each clock. Write the time.

7. half past 4

8. half past 7

9. half past 6

Problem Solving

10. Eli woke up at half past 7. He needs to leave for school one hour later. Show and write the time he needs to leave for school.

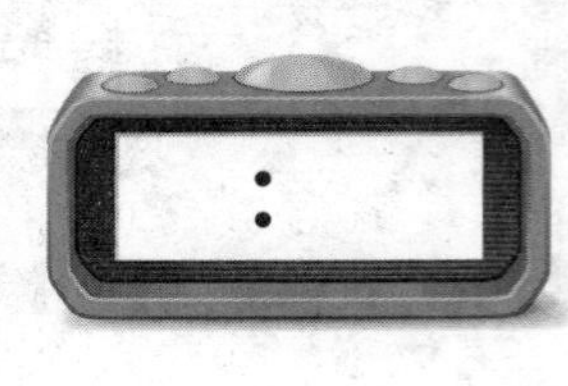

Brain Builders

Use the clocks shown below.

School starts

Soccer practice

Alarm goes off

What happens at each of these times?

11. two hours after half past 4 ____________________

12. one hour after half past 2 ____________________

13. one hour before half past 9 ____________________

Write Math How is reading a digital clock different from reading an analog clock? Explain.

____

____

____

____

Name

My Homework

Lesson 2

Time to the Half Hour

Homework Helper

Need help? connectED.mcgraw-hill.com

You can tell and write time to the half hour.

It is half past 3 or 3:30.

Practice

Tell what time is shown. Write the time.

1.

half past ______

2.

half past ______

3.

half past ______

Draw the hands on each clock. Write the time.

4. half past 1

5. half past 11

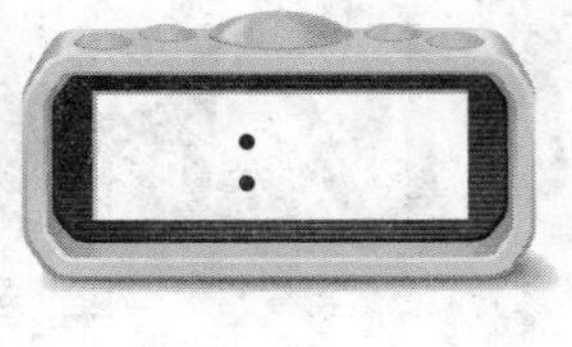

Brain Builders

6. Tatum is going to the city with her family. They will leave one hour after she gets home from school. Tatum gets home from school 2 hours after 1:30. What will the clocks look like when they leave for the city?

Vocabulary Check

Show a **half hour** later on the second clock.

7.

Math at Home Give your child a time to the hour. Have him or her tell you the position of the clock hands for half past that hour or in a half hour.

Name

Lesson 3

Problem Solving

STRATEGY: Find a Pattern

ESSENTIAL QUESTION
How do I use and tell time?

The buses leave in order every half hour. Bus 1 leaves at 9:30. Bus 2 leaves at 10:00. What time do buses 3 and 4 leave?

1 Understand

Underline what you know.
Circle what you need to find.

2 Plan

How will I solve the problem?

3 Solve

Find a pattern.

9:30	10:00	10:30	11:00
Bus 1	Bus 2	Bus 3	Bus 4

4 Check

Is my answer reasonable?

Practice the Strategy

Each class at school will go to recess 30 minutes after the other. The first class goes to recess at 1:00. What time will the fourth class go?

1 Understand Underline what you know. Circle what you need to find.

2 Plan How will I solve the problem?

3 Solve I will...

________, ________, ________, ________

Class 1 Class 2 Class 3 Class 4

4 Check Is my answer reasonable? Explain.

Name ______________________

Processes & Practices

Apply the Strategy

Find a pattern to solve. Use the space provided.

1. In the morning, Ms. White's students change activities every hour. Reading starts at 8:30. Her students go to three more activities. When do each of the activities start?

8:30, __________, __________, __________

Brain Builders

2. Students start work in learning stations at 10:30. After two hours, they go to lunch. Three hours after lunch begins they go home. What time do the students go home?

____

3. In the afternoon, Ms. White's students change activities every half hour. Math starts at 1:00. Writing starts after math. Reading starts after writing. What time will reading start?

____

Review the Strategies

Choose a strategy

- **Find a pattern.**
- **Act it out.**
- **Draw a picture.**

4. Carmen has 8 shoes. She places an equal number of shoes in two suitcases. How many shoes will be in each suitcase?

_____ shoes

5. A postcard costs 45¢. Jason wants to buy 2 postcards. He has 2 quarters, 2 dimes, and 3 nickels. How much money does he have?

____

How many pennies does he need so that he can buy the postcards?

_____ pennies

6. The subway arrives at the station on the hour and half hour. It is now 4:30. Kate's family is on the way but it will take one hour to get there. What is the earliest time her family can get on the subway?

____

Name

My Homework

Lesson 3

Problem-Solving: Find a Pattern

Ashley's family is going to visit their grandma. It will take them four hours to get there. They leave at 10:30. What time will they get to grandma's house?

1 Understand Underline what you know. Circle what you need to find.

2 Plan How will I solve the problem?

3 Solve Find a pattern.

10:30 → 11:30 → 12:30 → 1:30 → 2:30

1 hour, 1 hour, 1 hour, 1 hour

They will arrive at 2:30.

4 Check Is my answer reasonable?

Problem Solving

Underline what you know. Circle what you need to find. Find a pattern to solve.

1. There is one plane that leaves from the city to the ski resort every hour. The first plane leaves at 10:00. The last plane leaves at 4:00. How many planes fly to the resort each day?

_______ planes

Brain Builders

2. Mr. Lyon's class left for the zoo at 9:00. They arrived at the zoo in 30 minutes. They met the zookeeper 30 minutes later. Another 30 minutes after that, they saw the penguins. What time did they see the penguins?

____

3. Julio's family wants to take a ride on a tour bus. It leaves the bus station every half hour. It is 1:00. They just missed the bus. What are the times for the next two buses? Explain to a friend or family member how you solved the problem.

_______, _______

Math at Home Ask your child to keep a journal for one evening. At each half hour, have your child write the time and record his or her activity at that time. At the end of the evening, see if there were any repeating or other patterns in their activities.

Name ______________________

Check My Progress

Vocabulary Check 

Complete each sentence.

minute **analog clock** **hour**

digital clock **half hour**

1. 60 minutes is one ______________.
2. 30 minutes is one ______________.
3. 60 seconds is one ______________.
4. A ______________ uses only numbers to show time.
5. An ______________ has an hour hand and a minute hand to show time.

Concept Check

Tell what time is shown. Write the time.

6.

7.

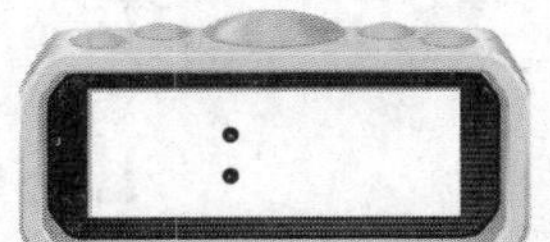

8.

Draw the hands on each clock. Write the time.

9. half past 8

:

10. 11 o'clock

:

11. half past 1

:

Write the time.

12.

half past ________

13.

________ o'clock

Brain Builders

14. Kevin has 3 classes in a row. The first two classes are each 2 hours long. His third class is 1 and a half hours long. Kevin's first class is at 7:00. When will Kevin's last class end?

____

15. **Test Practice** The class bell rings every 30 minutes. The first one rings at 8:30. What time will the 4th bell ring?

10:30	10:00	11:30	9:30
○	○	○	○

Name ____________________

Lesson 4
Time to the Quarter Hour

ESSENTIAL QUESTION
How do I use and tell time?

Math in My World

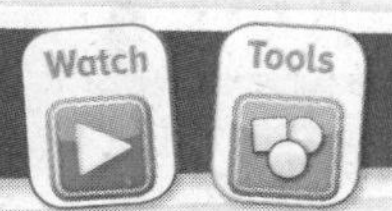

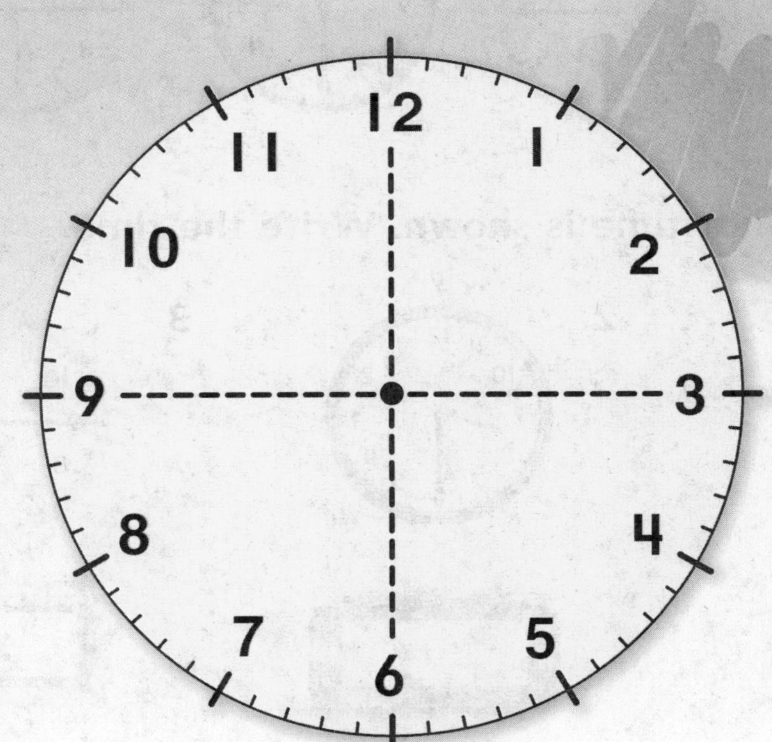

quarter past ____________

Teacher Directions: Have students fold a paper plate, or circular piece of paper in half. Fold it again in quarters. Unfold and look at the fold marks. On the clock above, place cubes on the quarter hours referring back to their plate or circle. Draw hands on the clock to show quarter past an hour. Write the time.

Processes &Practices

Guided Practice

You can show time to the **quarter hour**. There are 15 minutes in a quarter hour. There are 4 quarter hours in an hour.

Use [clock]. Tell what time is shown. Write the time.

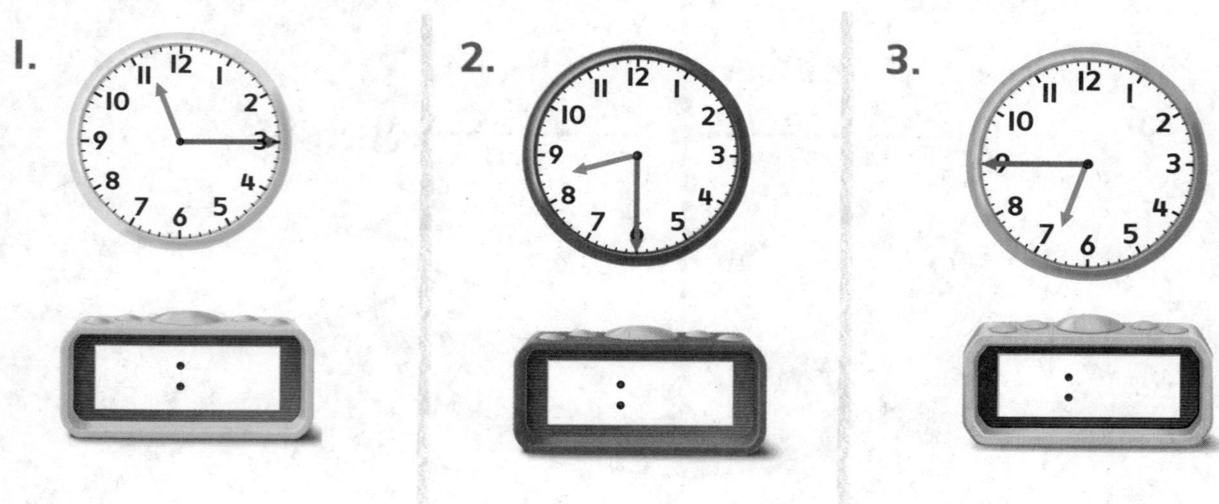

Talk Math At 4:15, where is the minute hand? Explain.

Name

Independent Practice

Use . Tell what time is shown. Write the time.

4. 

5. 

6. 

Draw the hands on each clock. Write the time.

7. quarter till 2

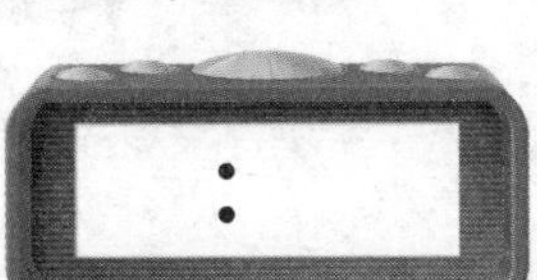

8. quarter past 5

9. quarter past 12

Read the time. Write the time on the digital clock.

10. twelve forty-five

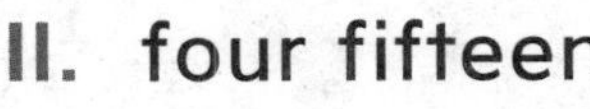

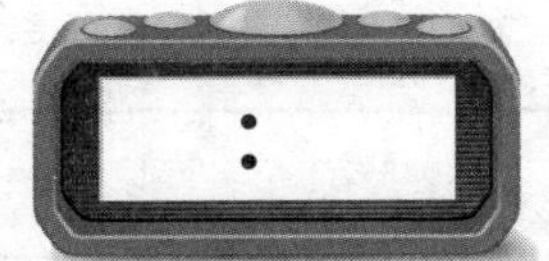

11. four fifteen

12. nine thirty

Problem Solving

13. Mia is at school. School is over at 3:15. She has 3 more hours to wait. What time is it now?

____

14. Ken goes to his friend's house at 2:15. Circle the clock that shows this time.

Brain Builders

15. Alea's family went hiking. They left one hour before quarter to 11. They drove for 3 hours. Then they stopped for lunch. Lunch took one hour. What time did they finish lunch?

____

16. Why is each 15-minute period on a clock called one quarter hour?

____

____

____

Name

My Homework

Lesson 4

Time to the Quarter Hour

Homework Helper

Need help? connectED.mcgraw-hill.com

You can show time to the quarter hour.

quarter past 8

half past 8

quarter till 9

Practice

Use . Tell what time is shown. Write the time.

1.

2.

3.

Draw the hands on each clock. Write the time.

4. quarter past 6 5. quarter till 3 6. quarter past 7

Brain Builders

7. Carly went to school at 8:15. She left school at 2:45. If she arrived at school at 8:45, how many hours was she in school?

_____ hours

Vocabulary Check

8. Circle the clock that shows an example of quarter hour.

Math at Home Have your child use the words quarter till and quarter past to describe the time at 6:15 and 6:45.

Name

Lesson 5
Time to Five Minute Intervals

ESSENTIAL QUESTION
How do I use and tell time?

Math in My World

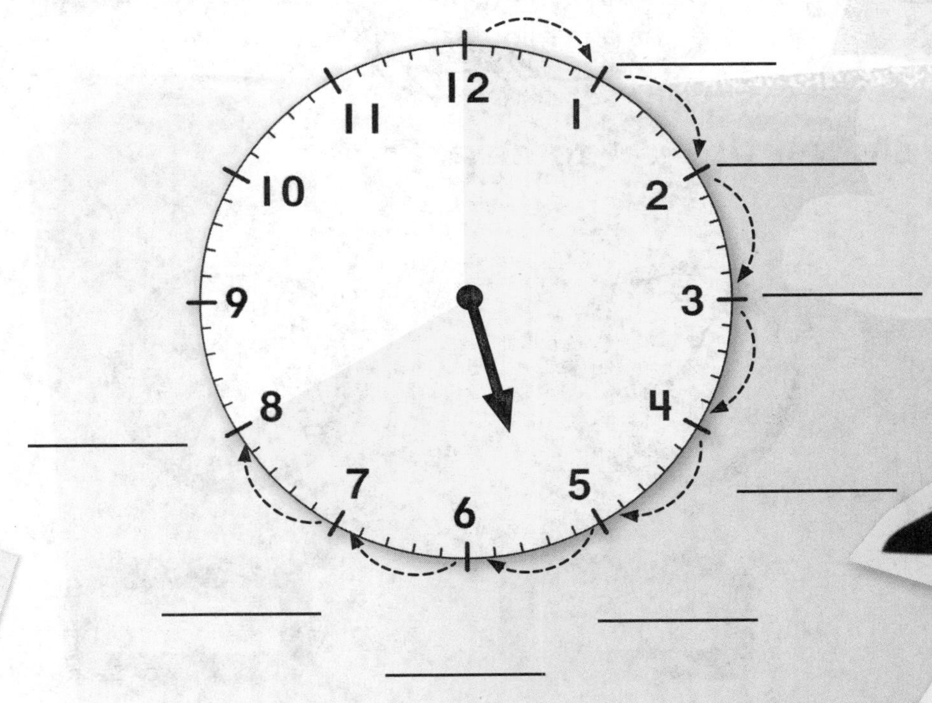

Teacher Directions: Count by 5s as you trace the dotted line. Label each jump 5, 10, 15 and so on. Draw the minute hand. Write that time on the digital clock.

Processes & Practices

Guided Practice

It takes 5 minutes for the minute hand to move to the next number. You can skip count by 5s to tell the time.

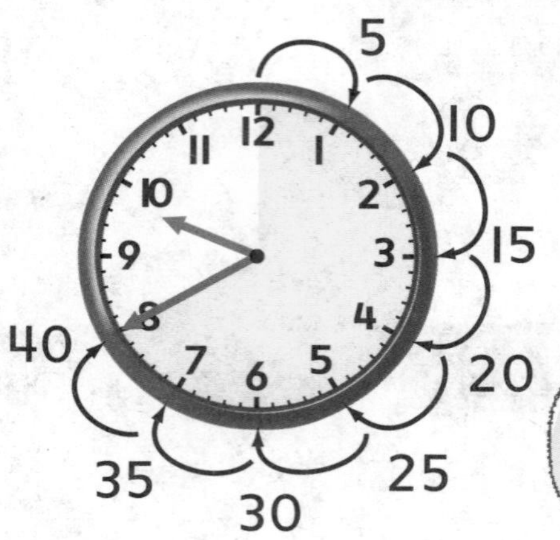

This clock shows 40 minutes after 9 o'clock. Write the time another way.

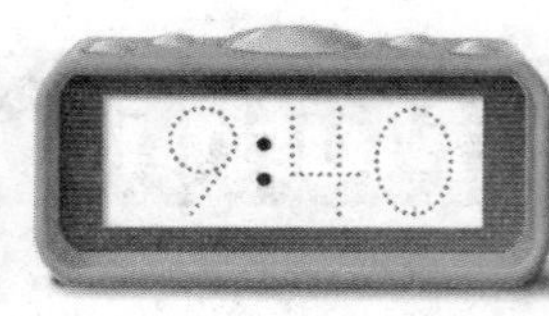

Tell what time is shown. Use [clock] to help. Write the time.

1.

2.

3.

4. Tell what time is shown. Draw the minute hand to show the time.

Talk Math Explain how you skip count by 5s to tell time.

Name

Independent Practice

Tell what time is shown. Use [clock] to help. Write the time.

5.

6.

7.

Tell what time is shown. Draw the hands to show the time.

8.

9.

10.

11. 2:20

12. 8:45

13. 1:25

Problem Solving

Processes & Practices

14. If the hour hand is close to the 11 and the minute hand is pointing to the 10. What time is it?

____

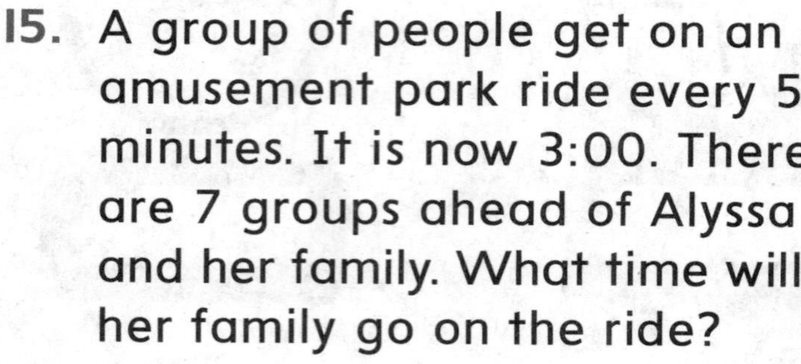

15. A group of people get on an amusement park ride every 5 minutes. It is now 3:00. There are 7 groups ahead of Alyssa and her family. What time will her family go on the ride?

Brain Builders

16. Will gets to the theater at 7 o'clock. It takes 20 minutes to get snacks and 5 minutes to find a seat. What time was Will ready to watch the movie?

Write Math What time is it when the hour hand is between 5 and 6 and the minute hand is on 7? Explain.

____

____

____

Name

My Homework

Lesson 5

Time to Five-Minute Intervals

Homework Helper

Need help? connectED.mcgraw-hill.com

It takes 5 minutes for the minute hand to move to the next number. Skip count by 5s to tell time.

This clock shows 35 minutes after 4 o'clock.

The time is shown another way.

Practice

Read the time. Write the time.

1.

2.

3.

Tell what time is shown. Draw the minute hand to show the time.

4. 

5. 

6. 

Brain Builders

7. It is 1:00. Hunter is waiting for Ben. Ben said he would meet Hunter in 25 minutes but he arrived 5 minutes late. What time does Ben meet Hunter?

___ : ___

8. **Test Practice** A train leaves at 10:40. Which clock shows 10:40?

○ ○ ○ ○

Math at Home Ask your child to look at an analog clock on the hour. Have him or her tell you what time it will be in 5 minutes, 10 minutes, 25 minutes, and 50 minutes.

Name

Lesson 6
A.M. and P.M.

ESSENTIAL QUESTION

How do I use and tell time?

Math in My World

Teacher Directions: Think of things that you do during the day and at night. Draw pictures of the activities on the correct side of the page.

Processes &Practices

Guided Practice

The hours from midnight until noon are labeled **A.M.** The hours from noon until midnight are labeled **P.M.**

Helpful Hint
12:00 P.M. is noon.
12:00 A.M. is midnight.

Go to school

8:00 A.M.

Read a bedtime story

8:00 P.M.

Tell what time is shown for the activity. Write the time. Circle A.M. or P.M.

1. Art class

A.M.
P.M.

2. Go to bed

A.M.
P.M.

3. Play after school

A.M.
P.M.

Talk Math How can you remember if it is A.M. or P.M.?

Name

Independent Practice

Tell what time is shown for the activity. Write the time. Circle A.M. or P.M.

4. Eat breakfast

A.M.
P.M.

5. Wash the dog

A.M.
P.M.

6. Go swimming

A.M.
P.M.

Tell what time is shown for the activity. Draw the hands on the clock to show the time. Circle A.M. or P.M.

7. Soccer practice

A.M.
P.M.

8. Sleeping in bed

A.M.
P.M.

9. Eat dinner

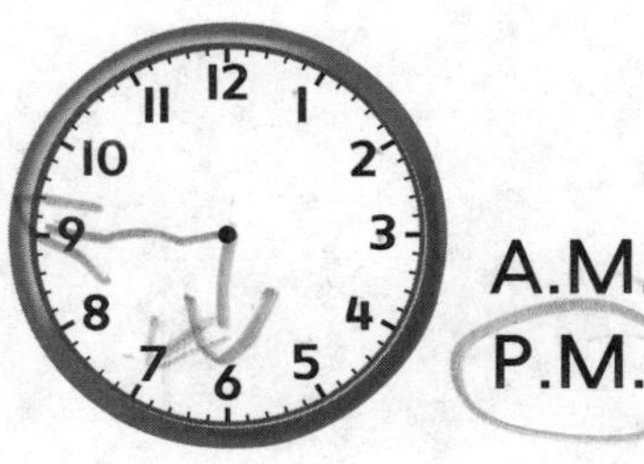

A.M.
P.M.

Complete each sentence. Write *midnight* or *noon*.

midnight **noon**

10. At Midnight, I am usually asleep.

11. At noon, I may be eating my lunch.

Problem Solving

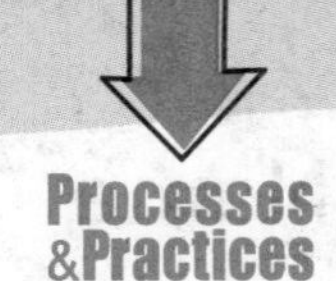

12. Zack has lacrosse practice at 4:00. Would that likely be at 4:00 A.M. or 4:00 P.M.?

P.M.

13. It is 2:25. Lakota is going to the library in 1 hour. On the clock show the time she will go to the library.
Circle A.M. or P.M.

A.M.  P.M.

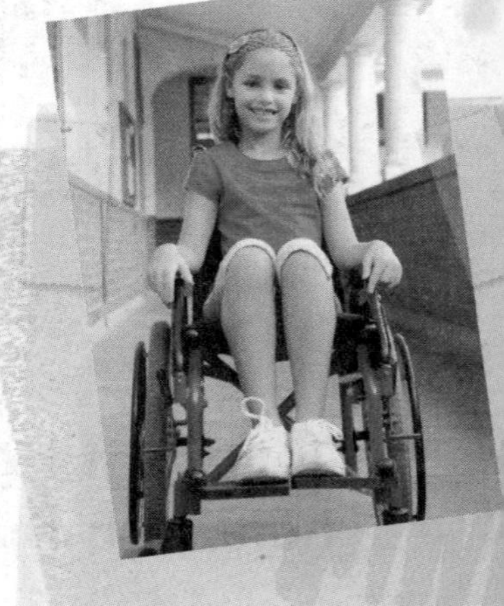

Brain Builders

14. It is 3:00 A.M. Are you more likely to be on your way home from school or sleeping in your bed? Explain to a friend how you know.

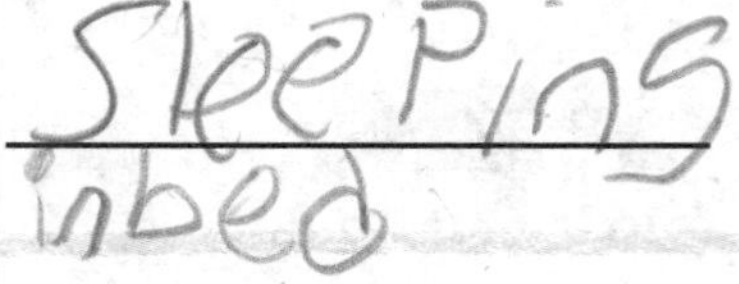

Write Math Christy wants to go to her friend Jill's party. The invitation says the party starts at 1:30 A.M. This is wrong. What mistake was made on the invitation?

iT say's it Starts at 1:30 A.m. you need 1:30 P.m

Name ....

My Homework

Lesson 6

A.M. and P.M.

Homework Helper

Need help? connectED.mcgraw-hill.com

The hours from midnight until noon are labeled A.M. The hours from noon until midnight are labeled P.M.

Helpful Hint
12:00 P.M. is noon.
12:00 A.M. is midnight.

Wake up

9:00 A.M.

Look at the moon

9:00 P.M.

Tell what time is shown for the activity. Write the time. Circle A.M or P.M.

1. Go to the park

A.M.
P.M.

2. Go bowling

A.M.
P.M.

3. Make your bed

A.M
P.M.

Tell what time is shown for the activity.
Draw the hands on the clock. Circle A.M. or P.M.

4. Rooster crowing

A.M.
P.M.

5. Flying a kite

A.M.
P.M.

6. Going shopping

A.M.
P.M.

Brain Builders

7. Jay and Mike got to the county fair at 10:00 A.M. They stayed for 4 and a half hours. Did they get home in the A.M. or P.M.? Exlain your reasoning to a friend or family member. P.m

Vocabulary Check

Vocab

Complete each sentence.

A.M. **P.M.**

8. Casey has an art lesson at 4:30 P.m.

9. Sam eats dinner at 5:30 P.m.

10. Kaylee eats breakfast at 6:00 A.m.

Math at Home Several times in the next 24 hours, ask your child for the time and then if it is A.M. or P.M.

Name ....

My Review

Chapter 10
Time

Vocabulary Check 

Complete each sentence.

hour	**half hour**	**quarter hour**	**minute**
A.M.	**P.M.**	**digital**	**analog**

1. The long hand on an analog clock is the Minute hand.

2. The short hand on an analog clock is the hour hand.

3. 15 minutes is a quarterhour.

4. 30 minutes is a half hour.

5. A clock that uses hands to show the time is an analog clock.

6. A clock that uses numbers to show the time is a digital clock.

7. The hours from noon to midnight are P.m..

8. The hours from midnight to noon are A.m..

Concept Check

Read the time. Write the time.

9. 

10. 

11. 

Tell what time is shown. Draw the hands on each clock.

12. 

13. 

14. 

Circle the best choice.

15.	Walk the dog.	11:45	A.M.	P.M.
16.	Go hiking.	2:30	A.M.	P.M.
17.	Wash dinner dishes.	6:30	A.M.	P.M.

Name

Date

Score

Performance Task

Making Time

Today is the school Science Fair where you will present your science project. The Science Fair is not until the evening. You have a busy day! The clocks below show your schedule.

School	Swimming	Science Fair
 (clock)	 (clock)	 (clock)

Show all your work to receive full credit.

Part A

Write each of the times shown on the clocks. Which times do you think are A.M.? Which times do you think are P.M.?

Part B

If you want to have time to play on the playground before school starts, you must arrive 15 minutes early. What time do you arrive to have 15 minutes to play on the playground? Write your answer and show the time on the clock below.

Part C

How much time do you have from when swimming starts to when the Science Fair starts? Explain how you know.

Part D

You arrive home from swimming, take a shower, and eat your dinner. It is now 6 P.M. How much time do you have before the Science Fair starts? Explain.

Chapter 11 Customary and Metric Lengths

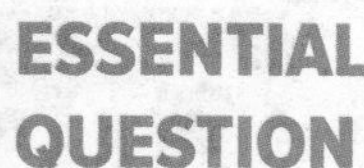

ESSENTIAL QUESTION

How can I measure objects?

Watch a video!

Name

Chapter 11 Project

Measurements All Around

1. Select various objects in the classroom to measure.
2. List each object and your estimate of the measure of the object. Then measure the object and record the measurement and the unit you measured in.
3. After you have measured at least 5 objects in each unit (inches, feet, yards, centimeters, and meters), list the objects you measured in each unit in order from smallest to largest.
4. Compare the items you measured with your classmates.
5. In the table below, enter real-world objects from outside the classroom that you might measure using each of the units you studied in this chapter, and the tool you might measure them with.

Unit	Object	Tool
inches		
feet		
yards		
centimeters		
meters		

Name ______________________

Am I Ready?

Add or subtract.

1. $\begin{array}{r} 34 \\ +18 \\ \hline \end{array}$

2. $\begin{array}{r} 26 \\ -17 \\ \hline \end{array}$

3. $\begin{array}{r} 38 \\ +36 \\ \hline \end{array}$

Measure the length of the object in cubes.

4.

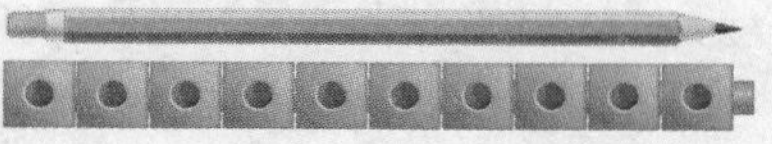

_______ cubes

5.

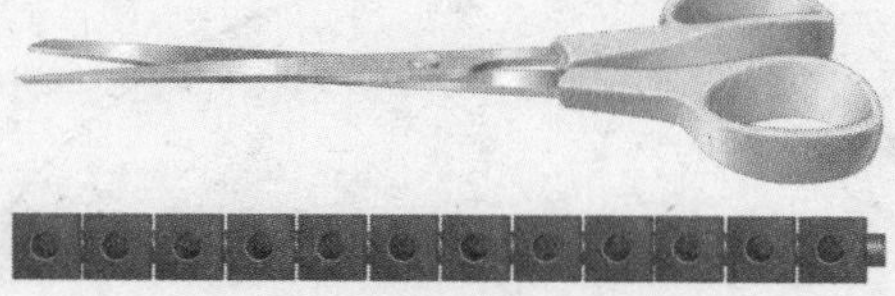

_______ cubes

6. Neil's hand is 5 cubes long. He measures this checkerboard. It is 3 hands long. How many cubes long is the checkerboard?

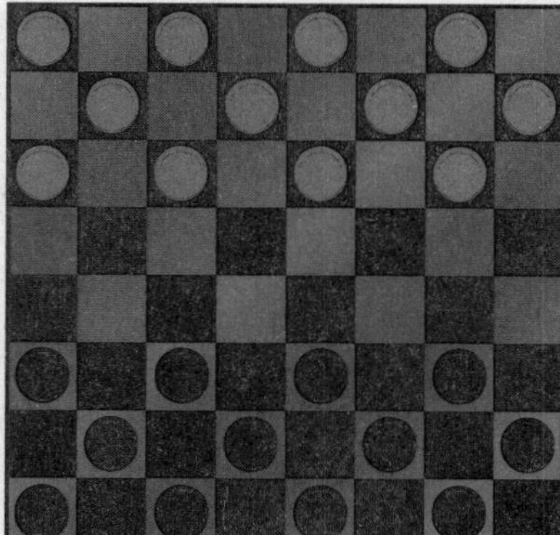

_______ cubes

How Did I Do?

Shade the boxes to show the problems you answered correctly.

1	2	3	4	5	6

Name ....

My Math Words

Vocab

Review Vocabulary

compare | longest | shortest

Find three classroom objects. List them. Compare their lengths. Write the longest object. Write the shortest object.

_______________ _______________ _______________

Longest **Shortest**

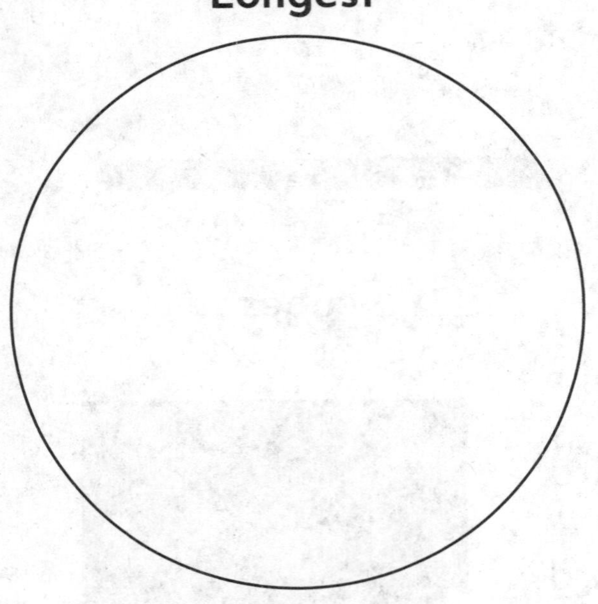

How did you compare your objects?

____

____

____

My Vocabulary Cards

Lesson 11-7

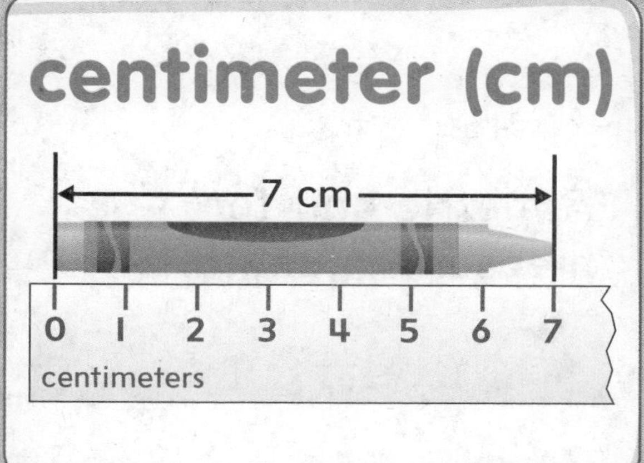

Lesson 11-1

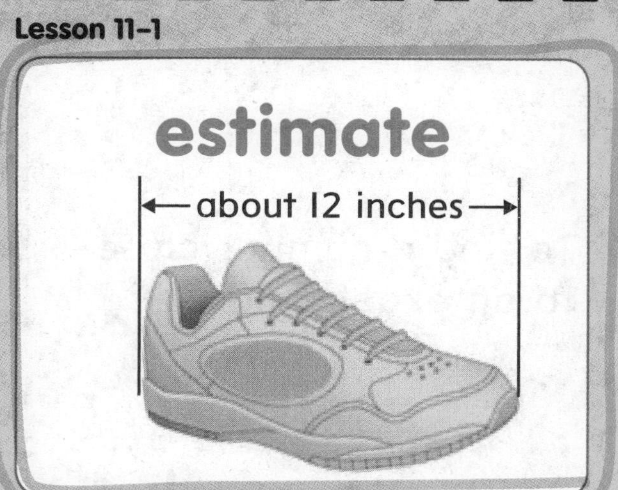

Lesson 11-2

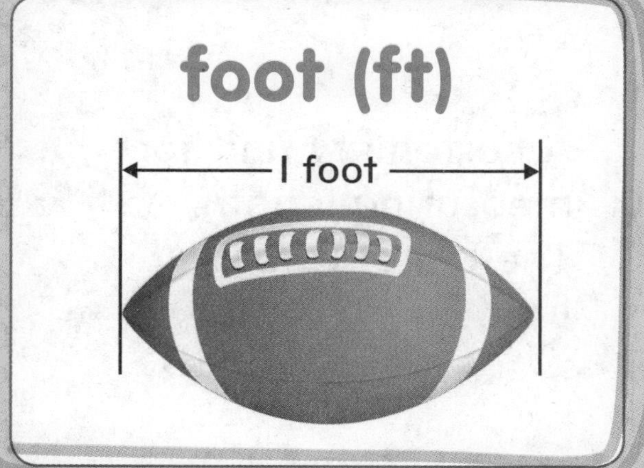

Lesson 11-1

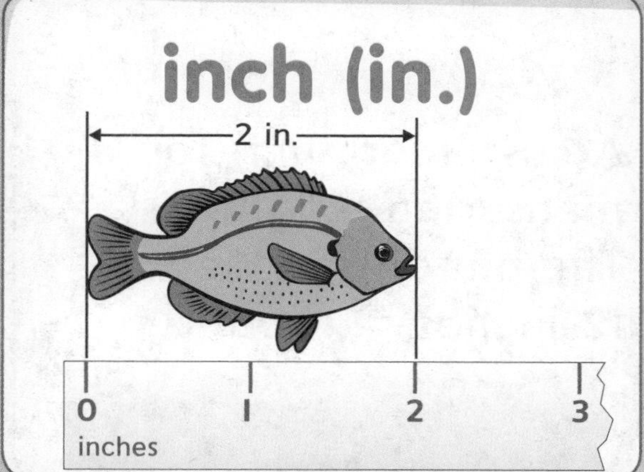

Lesson 11-1

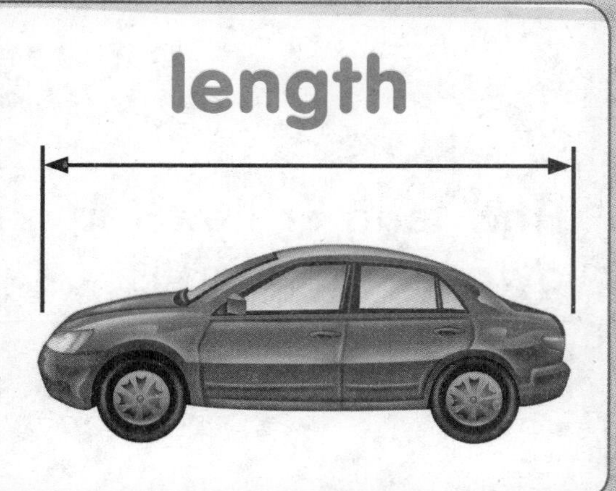

Lesson 11-1

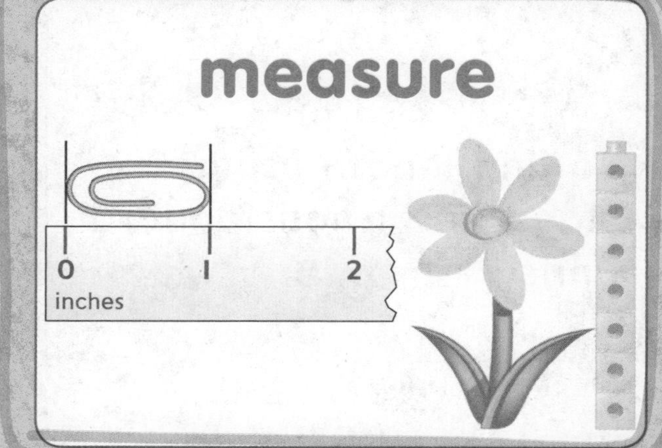

Directions:
Ideas for Use

- Have students group 2 or 3 common words. Ask them to add a word that is unrelated to the group. Have them ask a friend to name the unrelated word.
- Ask students to find pictures to show an example of each word. Have them ask a friend to guess which word the picture shows.

To find a number close to an exact amount.

A metric unit for measuring length.

A customary unit for measuring length. The plural is inches. 12 inches = 1 foot

A customary unit for measuring length. The plural is feet. 1 foot = 12 inches

To find length using standard or nonstandard units.

How long or how far away something is.

My Vocabulary Cards

Lesson 11-7

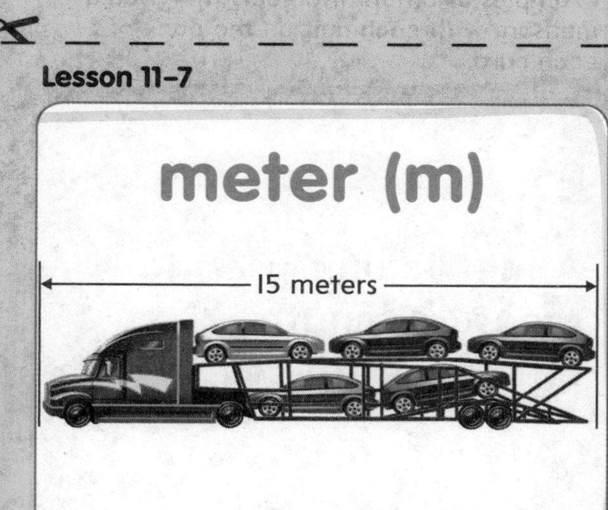

Lesson 11-2

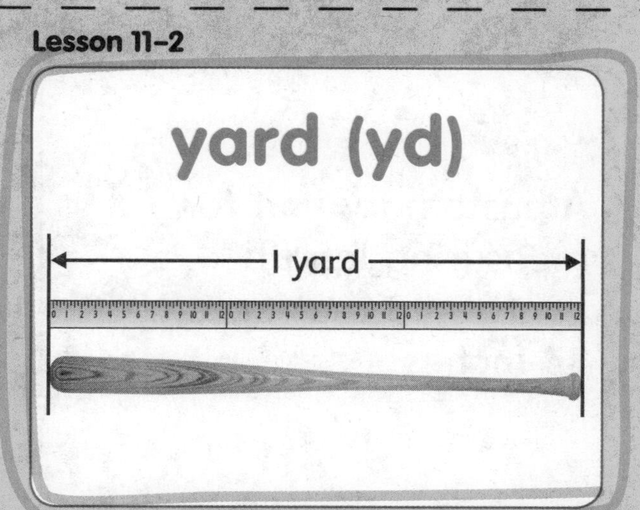

Directions:
More Ideas for Use

- Have students use a blank card to write this chapter's essential question. Have them use the back of the card to write or draw examples that help them answer the question.
- Ask students to use the blank cards to write some of the units of measure they learned in this chapter. They should give examples of different objects they would measure with each unit on the back of each card.

A customary unit for measuring length.
1 yard = 3 feet or 36 inches

A metric unit used to measure length.
1 meter = 100 centimeters

FOLDABLES® Follow the steps on the back to make your Foldable.

estimate

inches
yards

estimate

centimeters
meters

meters
centimeters

estimate

inches
feet

estimate

FOLDABLES®
Study Organizer

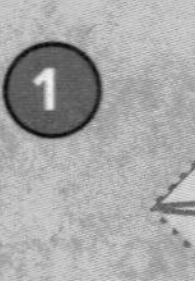

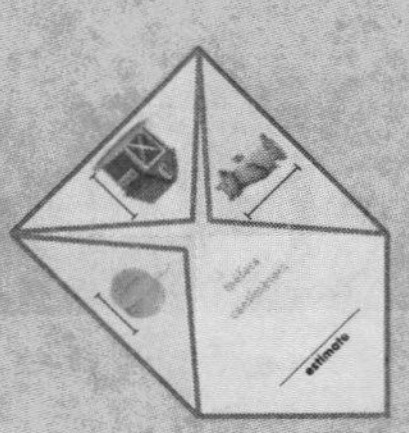

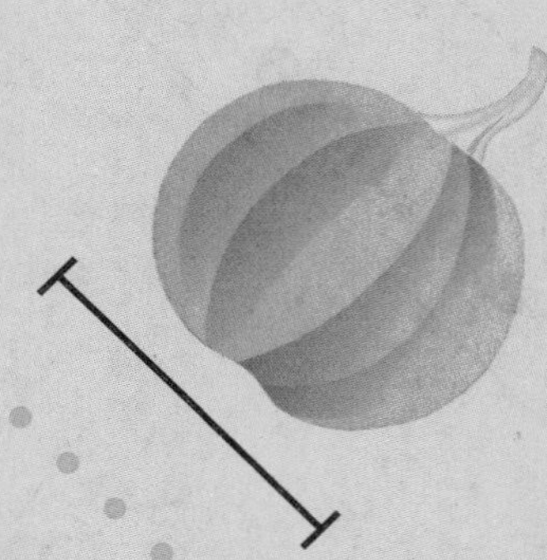

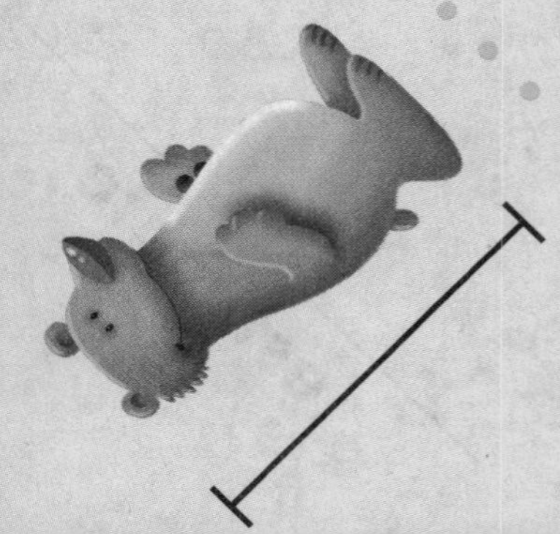

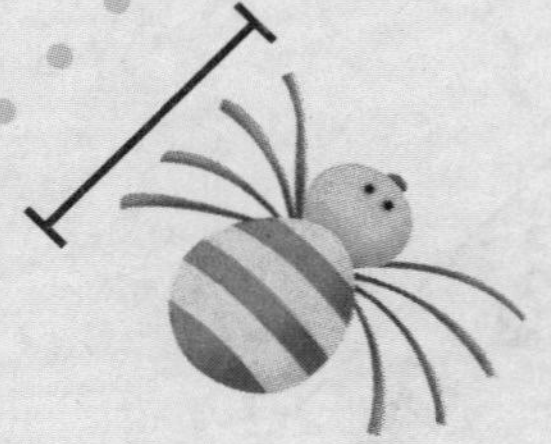

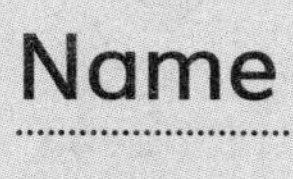

Name

Lesson 1
Inches

ESSENTIAL QUESTION
How can I measure objects?

Math in My World

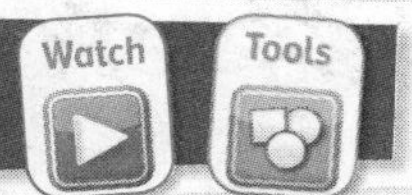

0 1 2 3
inches

_____ inches

_____ inches

_____ inches

_____ inches

Teacher Directions: Place a color tile above the ruler. Line it up with the 0. The color tile is one inch long. Use color tiles to measure the length of each object on the page. Write each length in inches.

Processes
&Practices

Guided Practice

One **inch** is about the **length** of one color tile. You can use what you know about inches to **estimate** the length of an object. Then use an inch ruler to **measure** the length.

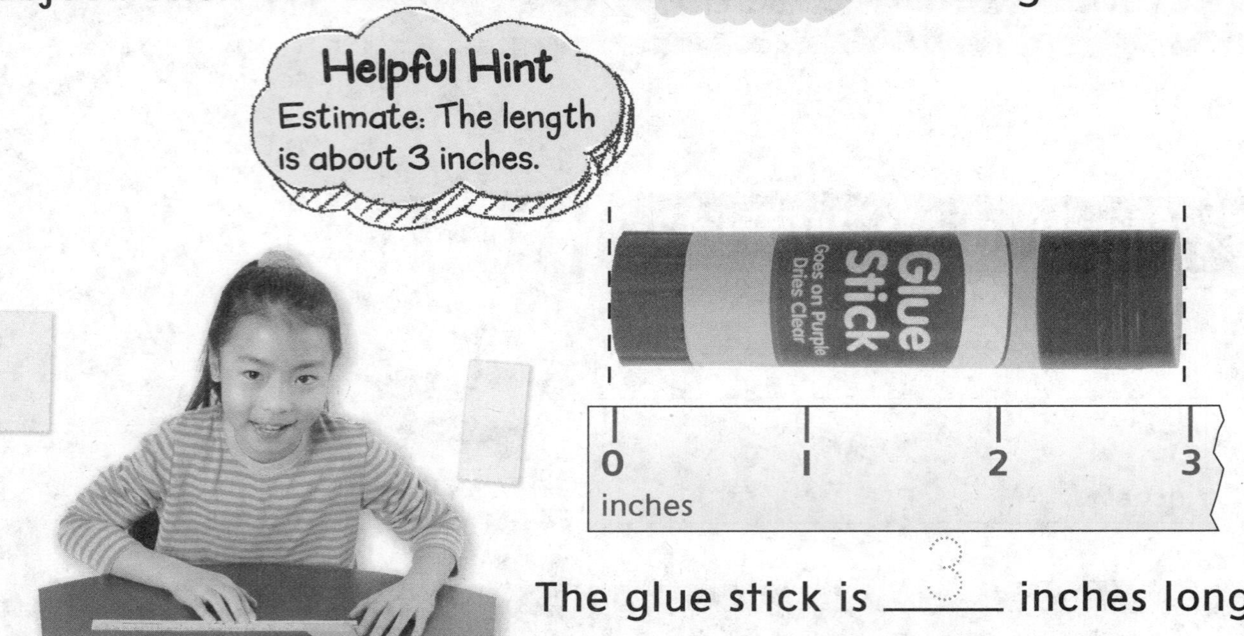

The glue stick is __3__ inches long.

Find the object. Estimate the length.
Measure each object in inches.

Object	Estimate	Measure
1.	about _____ inches	about _____ inches
2.	about _____ inches	about _____ inches

Talk Math How do you use a ruler to measure inches?

Name ....

Independent Practice

Find the object. Estimate the length. Measure each object in inches.

Object	Estimate	Measure
3.	about _____ inches	about _____ inches
4.	about _____ inches	about _____ inches
5.	about _____ inches	about _____ inches
6.	about _____ inches	about _____ inches
7.	about _____ inches	about _____ inches

Problem Solving

8. Kaya's surfboard must be at least 15 inches longer than 48 inches. Should Kaya choose a surfboard that is 55 inches long or 65 inches long?

______ inches long

9. Owen's foot is 6 inches long. His dad's foot is 6 inches longer than Owen's foot. How long is his dad's foot?

______ inches

Brain Builders

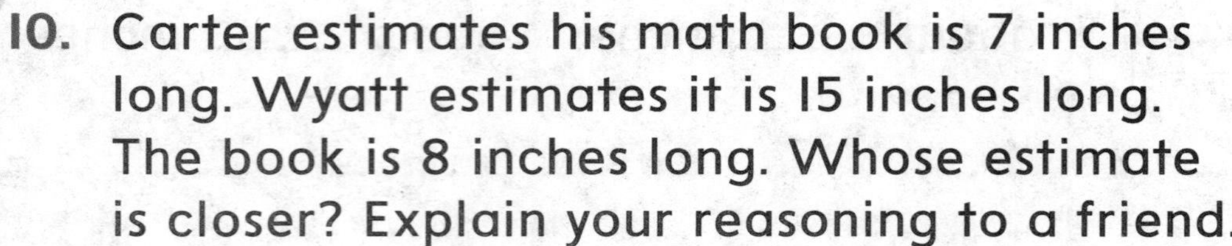

10. Carter estimates his math book is 7 inches long. Wyatt estimates it is 15 inches long. The book is 8 inches long. Whose estimate is closer? Explain your reasoning to a friend.

____

Write Math How are an estimate and an actual measurement different?

____

____

____

Name ______________________________

Lesson 1

Inches

My Homework

Homework Helper

Need help? connectED.mcgraw-hill.com

Estimate the length of an object. Then check your estimate by measuring the length with an inch ruler.

Helpful Hint
A small paper clip is about 1 inch long.

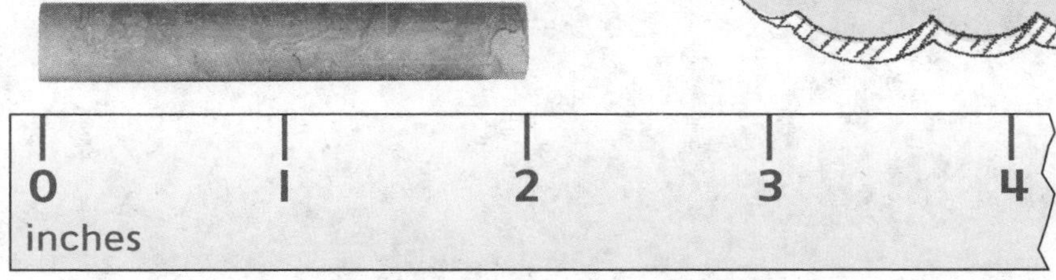

The chalk is 2 inches long.

Practice

Find the object. Estimate the length. Measure each object in inches.

Object	Estimate	Measure
1.	about ______ inches	about ______ inches
2.	about ______ inches	about ______ inches

Find two objects. Draw them. Estimate the length. Measure each object in inches.

Object	Estimate	Measure
3.	about ______ inches	about ______ inches
4.	about ______ inches	about ______ inches

Brain Builders

5. A red piece of yarn is 22 inches long. A green piece of yarn is 34 inches long. How much longer is the green piece of yarn? Write a number sentence that shows how you solved the problem.

______ inches

____

Vocabulary Check

Circle the correct answer.

estimate **inch** **measure** **length**

6. An ______ is a unit for measuring length.

Math at Home Ask your child to measure the length of a fork and a spoon using an inch ruler.

Name

Lesson 2
Feet and Yards

ESSENTIAL QUESTION
How can I measure objects?

Math in My World

EXIT

Teacher Directions: Discuss each object in the picture. Circle the objects that could be measured with an inch ruler. Draw an X on the objects that are too big to be measured with an inch ruler.

Processes &Practices

Guided Practice

You can measure in feet or yards. A **foot** is equal to 12 inches. A **yard** is equal to 36 inches. Length can be measured in any direction.

about 3 feet

about 1 yard

Find the object. Estimate the length. Measure each object in feet or yards.

Object	Estimate	Measure
1.	about ______ feet	about ______ feet
2.	about ______ yards	about ______ yards

Talk Math How can you measure a large object with a ruler?

Name

Independent Practice

Find the object. Estimate the length. Measure each object in feet or yards.

Object	Estimate	Measure
3. 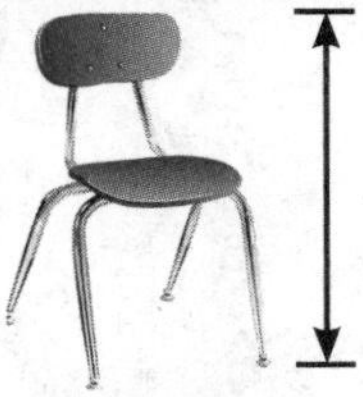	about ______ feet	about ______ feet
4. 	about ______ feet	about ______ feet
5. 	about ______ feet	about ______ feet
6. 	about ______ feet	about ______ feet
7. 	about ______ yard	about ______ yard

Problem Solving

8. The length of Sydney's tennis racket is 2 rulers long. How many feet long is Sydney's tennis racket?

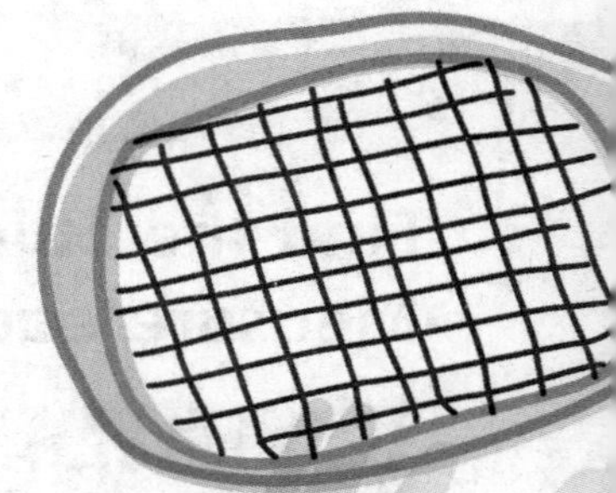

_______ feet

9. Gabriella is 2 inches taller than a yard. How many inches tall is Gabriella?

_______ inches

Brain Builders

10. The pool at the park is 19 feet wide. How many yardsticks would it take to measure the width of the pool? Explain.

____

____

11. Josh used 10 yardsticks to measure the length of his basement. How many rulers would he have used? Explain.

____

____

____

Name ....

Lesson 2

Feet and Yards

My Homework

Homework Helper 

Need help? connectED.mcgraw-hill.com

One foot is equal to 12 inches. One yard is equal to 36 inches. A ruler is 1 foot long. A yardstick is 1 yard long.

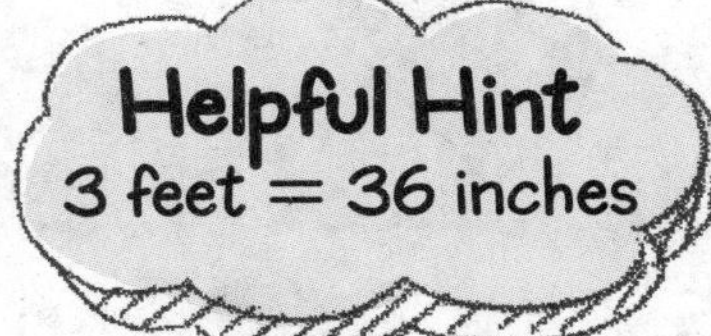

ruler

yardstick

Practice

Find the object. Estimate the length. Measure each object in feet or yards.

Object	Estimate	Measure
1.	about ______ feet	about ______ feet
2.	about ______ yards	about ______ yards

Find the object. Estimate the length. Measure each object in feet or yards.

Object	Estimate	Measure
3.	about ______ yards	about ______ yards
4.	about ______ foot	about ______ foot
5.	about ______ feet	about ______ feet

Brain Builders

6. The sidewalk from Ameila's house to her driveway is 96 inches long. How many feet long is the sidewalk?

______ feet

Sidewalk race!

Vocabulary Check 

Circle the correct answer.

7. **foot**

3 yards   12 inches   36 inches   1 yard

Math at Home Ask your child to measure his or her room using a yardstick or a ruler.

Name

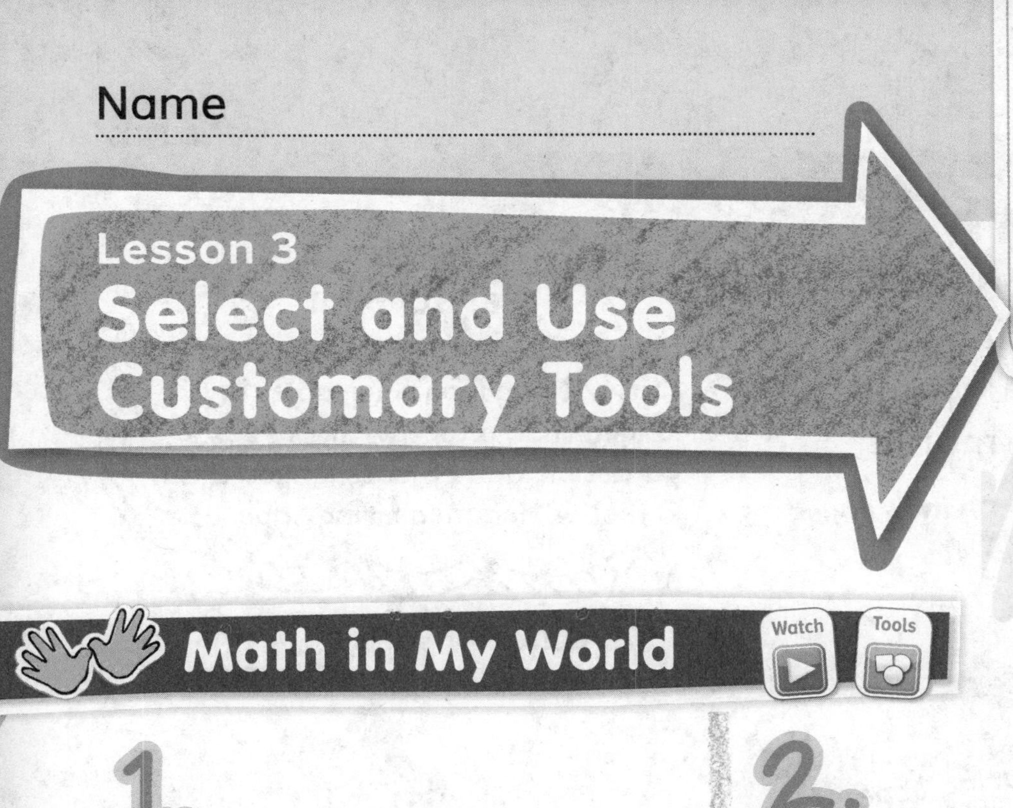

Lesson 3
Select and Use Customary Tools

ESSENTIAL QUESTION
How can I measure objects?

Math in My World

Watch

Tools

3.

Teacher Directions: Draw an object you would measure with an inch ruler in box 1. Draw an object you would measure with a yardstick in box 2. Draw an object you would measure with a measuring tape in box 3. Explain your drawings.

Processes
&Practices

Guided Practice

You can select and use tools to measure length. Circle the tool you would use to measure a marker.

Helpful Hint
Measure objects shorter than a foot with an inch ruler, objects longer than a foot with a yardstick, and objects longer than 3 feet with a measuring tape.

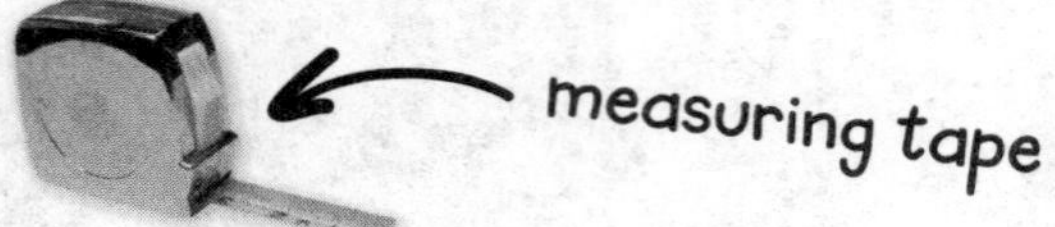

Find the object. Choose the tool and measure it. Explain why you chose that tool.

Object	Tool	Measure
1.	______	about ______
2.	______	about ______
3.	______	about ______

Talk Math How do you know which tool to use to measure?

Name ________________________________

Independent Practice

Find the object. Choose the tool and measure it. Explain why you chose that tool.

Object	Tool	Measure
4.	________	about ________
5.	________	about ________
6.	________	about ________
7.	________	about ________
8.	________	about ________
9.	________	about ________

Problem Solving

10. Vijay measures the length of his bike with a yardstick. He says it is one yardstick long. How many inches long is it?

_______ inches

11. Ethan measured his foot with a yardstick. What other tool would have been a better choice?

____

Brain Builders

12. Aisha measured her backyard with a ruler. Curtis measured his backyard with a yardstick. What other tool would have been a better choice? Explain your answer to a friend.

____

13. Omar laid 6 yardsticks end to end to measure his driveway. He said it was 6 feet long. Tell why Omar is wrong. Make it right.

____

____

____

Name ....

My Homework

Lesson 3

Select and Use Customary Tools

Homework Helper

Need help? connectED.mcgraw-hill.com

You can select and use tools to measure length.

inch ruler

yardstick

measuring tape

Helpful Hint

Measure objects shorter than a foot with an inch ruler, objects longer than a foot with a yardstick, and objects longer than 3 feet with a measuring tape.

Practice

Find the object. Choose the tool and measure it.

Object	Tool	Measure
1.	________	about ________
2.	________	about ________

Find the object. Choose the tool and measure it.

Object	Tool	Measure
3. 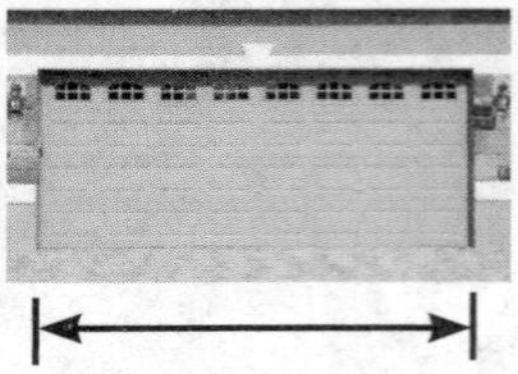	________	about ______
4. 	________	about ______
5. 	________	about ______

Brain Builders

6. There are 36 inches in 3 feet. How many inches are in 6 feet?

_____ inches

7. **Test Practice** Four feet is equal to _____ inches.

12	24	36	48
○	○	○	○

Math at Home Have your child measure the width of his or her room using a ruler, a yardstick, and a measuring tape. Talk about which tool is the best choice.

Name ....

Check My Progress

Vocabulary Check

Draw a line to match each word to the correct sentence.

1. **length**	To find a number close to an exact amount.
2. **inch**	Distance from end to end.
3. **estimate**	A customary unit for measuring length.

Concept Check

Find the object. Estimate the length.
Measure each object.

Object	Estimate	Measure
4.	about ______ inches	about ______ inches
5.	about ______ feet	about ______ feet
6.	about ______ yards	about ______ yards

Find the object. Choose the tool and measure it.

Object	Tool	Measure
7.	________	about ________
8. Lunch time	________	about ________
9.	________	about ________

Brain Builders

10. James measured the fence in his backyard with a ruler. It was 24 feet long. Which tool should James have used to measure the fence? Explain.

____

____

11. **Test Practice** Mr. Tom's driveway measures 7 yards long. How many feet long is Mr. Tom's driveway?

7	14	21	28
○	○	○	○

Name

Lesson 4
Compare Customary Lengths

ESSENTIAL QUESTION 

How can I measure objects?

Math in My World

Helpful Hint
Measure length up and down to measure how tall.

Teacher Directions: Find two objects in your classroom. Compare the lengths. Draw the objects. Explain which object is shorter and which object is longer.

Guided Practice

You can compare lengths of objects.

baseball bat

33 inches

golf club

31 inches

The baseball bat is 33 inches long.

The golf club is 31 inches long.

The golf club is 2 inches shorter than the bat.

Think:

$$\begin{array}{r} 33 \\ -31 \\ \hline 2 \end{array}$$

Find the objects. Measure them. Write the lengths. Write longer or shorter.

1. 

_______ inches _______ inches

The pencil is _______ inches ___________.

2.  

_______ feet _______ feet

The desk is _______ feet ___________.

Talk Math Why do you need to know how to compare lengths?

Name ____________

Independent Practice

Find the objects. Measure them. Write the lengths. Write longer or shorter.

3. 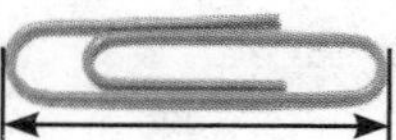 

_______ inches _______ inches

The paper clip is _______ inches _________.

4.  

_______ feet _______ feet

The car is _______ feet _____________.

5. 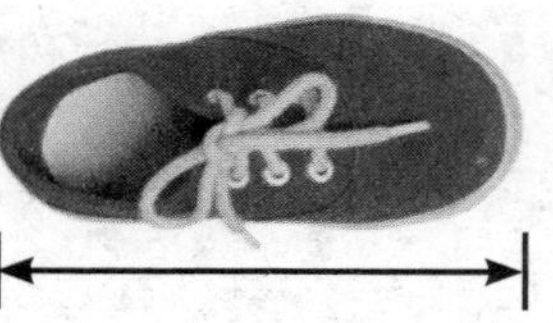 

_______ inches _______ inches

The shoe is _______ inches _____________.

6.  

_______ yards _______ yards

The door is _______ yards _____________.

Problem Solving

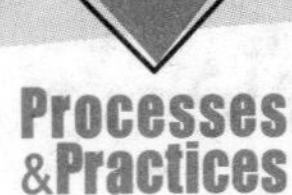

7. The playground is 90 yards long. The field by my house is 80 yards long. How much longer is the playground?

_______ yards

8. The blue hockey stick is 4 feet long. The red hockey stick is 6 feet long. How much longer is the red hockey stick?

_______ feet

Brain Builders

9. Audrey's basketball hoop is 7 feet tall. Isaac's basketball hoop is 1 yard taller. How tall is Isaac's basketball hoop?

_______ feet

10. Leah says her classroom is 7 yards wide. Austin says it is 21 feet wide. Both students are correct. Explain.

____

____

____

Name ___

My Homework

Lesson 4

Compare Customary Lengths

Homework Helper

Need help? connectED.mcgraw-hill.com

You can compare the lengths of objects.

72 inches

60 inches

The green jump rope is 12 inches longer.

Practice

Find the objects. Measure them.
Write the lengths. Write longer or shorter.

1.

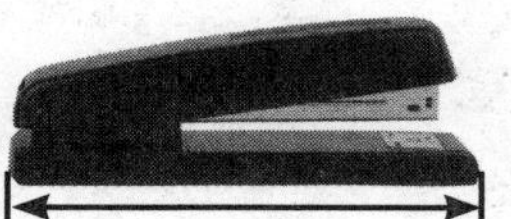

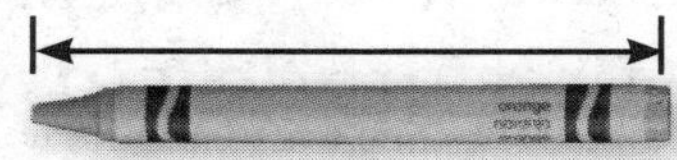

_______ inches _______ inches

The crayon is _______ inches __________.

2.

_______ feet _______ feet

The rug is _______ feet __________.

Find the objects. Measure them. Write the lengths. Write longer or shorter.

3. 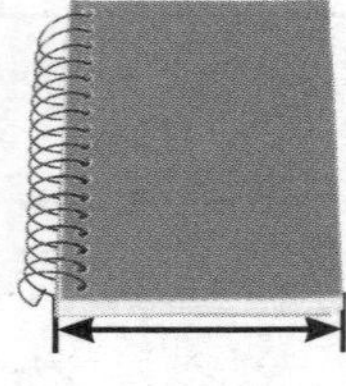

_______ inches _______ inches

The pencil is _______ inches __________.

4.  

_______ feet _______ feet

The window is _______ feet __________.

5. The second grade classroom is 24 feet wide. The first grade class is 3 feet wider. How wide is the first grade classroom?

____

Brain Builders

6. **Test Practice** 2 yards is equal to _______ feet.

3 ◯ 6 ◯ 24 ◯ 32 ◯

Math at Home Have your child measure two rooms in your house and compare the lengths.

Name ______________________

Lesson 5
Relate Inches, Feet, and Yards

ESSENTIAL QUESTION
How can I measure objects?

 Math in My World 

_______ inches or _______ feet

_______ inches or _______ feet

_______ inches or _______ feet

Teacher Directions: Find each object in the classroom. Measure each object in inches. Measure each object in feet. Discuss the measurements.

Guided Practice

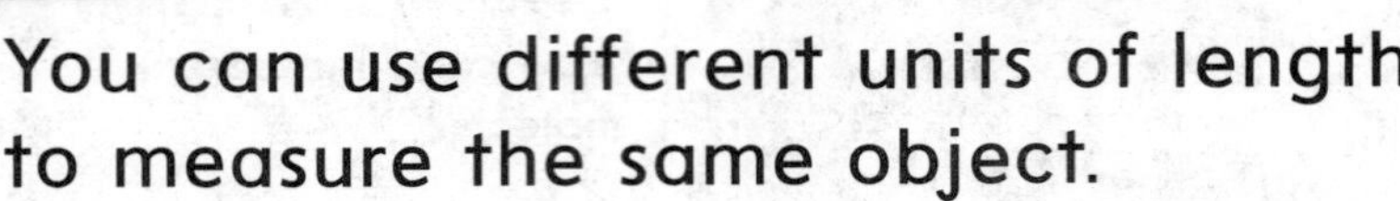

You can use different units of length to measure the same object.

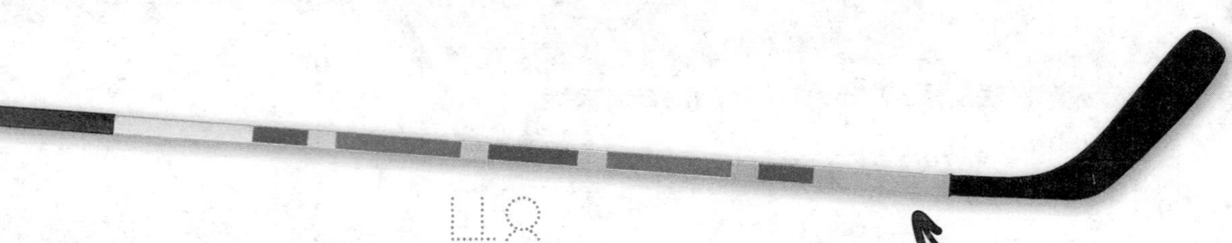

The hockey stick is ___48___ inches long.

It is ___4___ feet long.

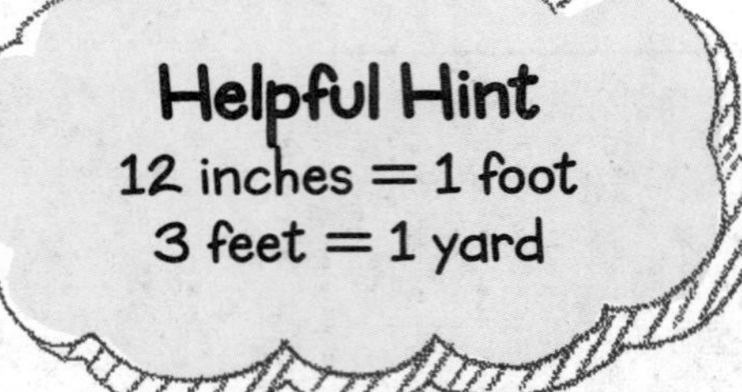

Find each object. Measure the length of each object twice.

Object	Measure
1.	______ feet ______ yards
2.	______ yards ______ feet

Talk Math If there are 12 inches in 1 foot and 3 feet in 1 yard, how many inches are in 1 yard?

Name

Independent Practice

Find each object. Measure the length of each object twice.

Object	Measure
3.	_______ feet _______ inches
4.	_______ inches _______ feet
5.	_______ feet _______ yards
6.	_______ yards _______ feet
7.	_______ feet _______ yards
8.	_______ inches _______ feet

Problem Solving

9. Sheradon's scooter is 3 feet long. Sage's scooter is 38 inches long. Whose scooter is longer?

____

10. The low diving board is 8 feet long. The high diving board is 4 yards long. How many feet longer is the high diving board?

_______________ feet

Brain Builders

11. The green bike is one yard long. The yellow bike is 38 inches long. Which bike is shorter? Explain your answer to a friend.

_______________ bike

Write Math Explain how the measurements of an object change depending on which unit you use to measure.

____

____

____

Name ....

My Homework

Lesson 5

Relate Inches, Feet, and Yards

Homework Helper

Need help? connectED.mcgraw-hill.com

You can use different units of length to measure the same object.

The rope is 21 feet long.

It is 7 yards long.

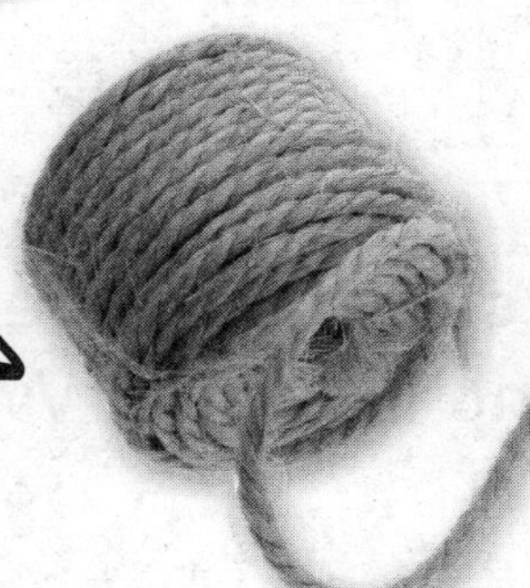

Practice

Find each object. Measure the length of each object twice.

Object	Measurement
1.	_______ feet _______ yards
2.	_______ yards _______ feet
3.	_______ inches _______ feet

Find each object. Measure the length of each object twice.

Object	Measurement
4.	_______ feet _______ inches
5.	_______ feet _______ yards

6. Sam's closet is 4 feet wide. Tom's closet is 50 inches wide. Whose closet is wider?

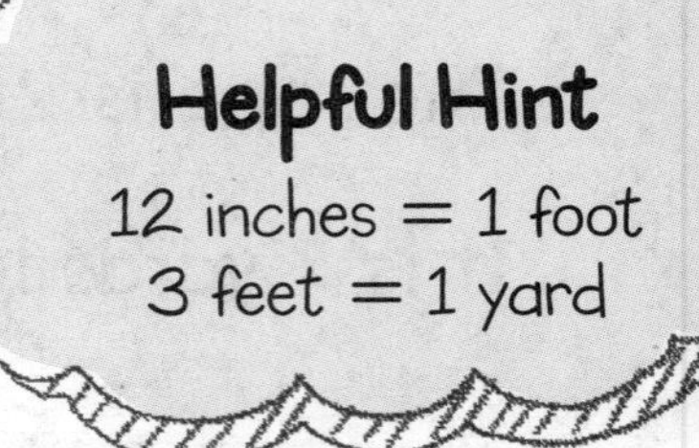

____

Brain Builders

7. The fence in Laura's backyard is 15 feet long. Is the fence longer than 6 yards? Explain.

____

____

8. **Test Practice** 72 inches is equal to _______ yards.

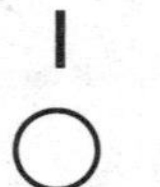

1	2	3	4
○	○	○	○

Math at Home Have your child measure an object in your house using inches, feet, and yards. Discuss how each measurement relates to the size of the unit.

Name

Lesson 6

Problem Solving

STRATEGY: Use Logical Reasoning

ESSENTIAL QUESTION

How can I measure objects?

Koko wants to plant a garden. She cannot decide if it should be 10 inches, 10 feet, or 100 yards long. About how long should the garden be?

1 Understand

Underline what you know.
Circle what you need to find.

2 Plan

How will I solve the problem?

3 Solve

Use logical reasoning.

10 inches is too short.

100 yards is too long.

The garden should be 10 feet long.

4 Check

Is my answer reasonable? Explain.

Practice the Strategy

Zachary is looking up at the high diving board at the pool. Is the high diving board 10 inches high, 10 feet high, or 10 yards high?

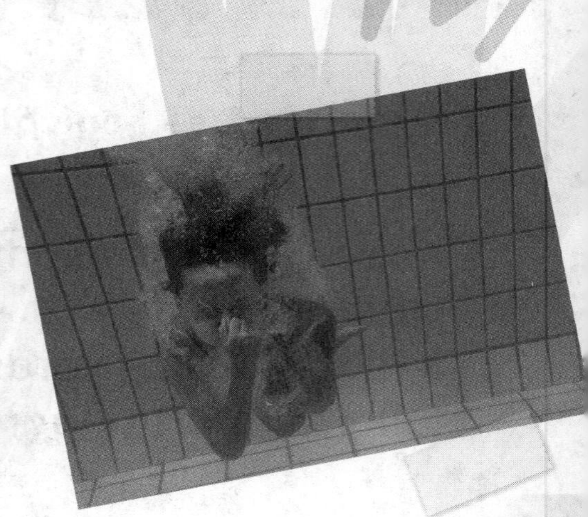

1 Understand Underline what you know.
Circle what you need to find.

2 Plan How will I solve the problem?

3 Solve I will...

4 Check Is my answer reasonable? Explain.

Name ....

Processes & Practices

Apply the Strategy

1. Sam planted a tomato plant that is 1 foot tall. The plant grows a little each week. After 4 weeks, would the plant be 10 inches or 14 inches tall?

_______ inches

Brain Builders

2. Jane made a paper chain that is 1 yard long. Brad made a paper chain that is 2 feet long. Who made the longer paper chain?

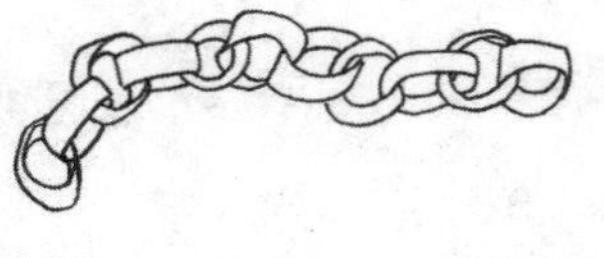

____

How many feet longer is it? ______ feet

3. Mr. Moore's class is collecting things to measure. Lisa finds a pinecone. Would the pinecone be 3 inches, 3 feet, or 3 yards long? Explain your answer to a friend.

____

Review the Strategies

Choose a strategy

- **Use logical reasoning.**
- **Write a number sentence.**
- **Make a model.**

4. Santos connects 6 markers. Each marker measures 5 inches. How long are all the markers together?

_____ inches

5. Jamil had a pencil that was 6 inches long. After a month of use it was only an inch long. How many inches did Jamil use?

_____ inches

6. Suni measured from one end of her kitchen table to the middle. The distance was 42 inches long. Then she measured from the middle to the other end. This distance measured the same length. How long is the table?

_____ inches

Name

My Homework

Lesson 6

Problem Solving: Use Logical Reasoning

Dave is more than 40 inches tall. He is less than 43 inches tall. His height is an even number. How tall is Dave?

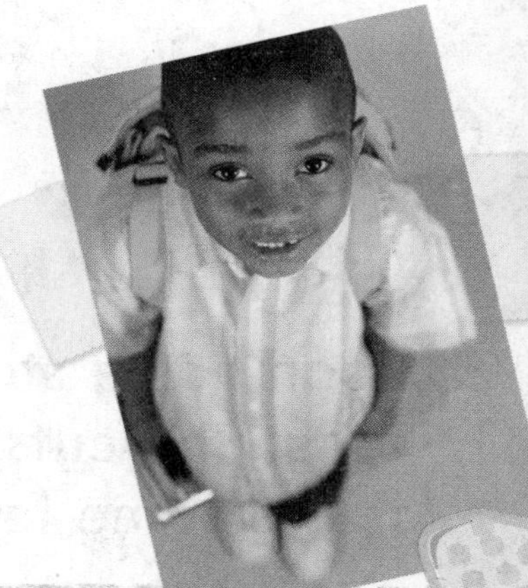

1 Understand Underline what you know. Circle what you need to find.

2 Plan How will I solve the problem?

3 Solve Use logical reasoning.
Dave is either 41 or 42 inches tall.
42 is an even number.
Dave is 42 inches tall.

4 Check Is my answer reasonable?

Problem Solving

Underline what you know. Circle what you need to find. Use logical reasoning to solve.

1. A bookshelf is 14 inches wide. There is enough space to fit 2 trophies. One trophy is 2 inches wider than the other. How wide is the wider trophy?

_______ inches

2. Joan has a stalk of celery that is 6 inches long. She cuts it into two pieces. Each piece is the same length. How long is each piece?

_______ inches

Brain Builders

3. Marlon and Brent are each measuring one of their shoes. Their shoes measure 15 inches in all. Marlon's shoe is 1 inch longer than Brent's shoe. How long is Brent's shoe?

_______ inches

4. **Test Practice** 36 inches is _______ feet.

1	2	3	4
○	○	○	○

Math at Home Ask your child to estimate distances from one room to another. Have your child check by measuring.

Name

Check My Progress

Vocabulary Check 

Draw lines to match.

1. measure — To find the length using standard or nonstandard units.
2. foot — Equal to 36 inches.
3. yard — Equal to 12 inches.

Concept Check

Find the objects. Measure them. Write the lengths. Write longer or shorter.

4. 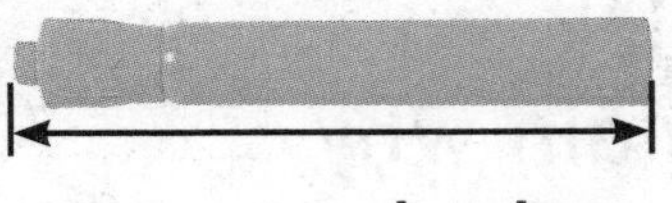

_______ inches

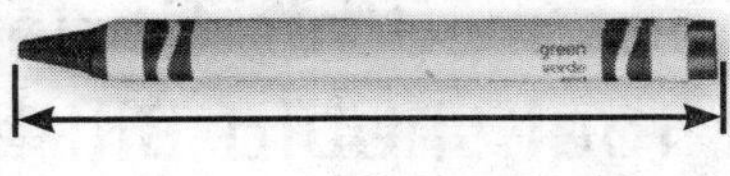

_______ inches

The marker is _______ inches _____________.

5. 

_______ feet

_______ feet

The cabinet is _______ feet _____________.

Find each object. Measure the length of each object twice.

Object	Measure
6.	_______ feet _______ yards
7.	_______ inches _______ feet
8.	_______ inches _______ feet

Brain Builders

9. Bill measured his finger with a yardstick. What tool should Bill have used? Explain why.

____

____

10. **Test Practice** Emily drew a blue line that is 1 foot long. She drew a purple line that is 4 inches long. How long are both lines in all?

○ 12 inches ○ 14 inches ○ 16 inches ○ 124 inches

Name

Lesson 7
Centimeters and Meters

ESSENTIAL QUESTION
How can I measure objects?

Math in My World

Watch

_____ centimeters

_____ centimeters

_____ centimeters

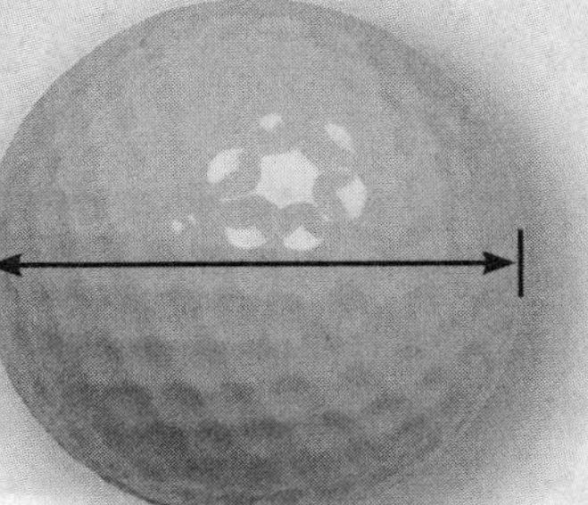

_____ centimeters

Teacher Directions: One unit cube is 1 centimeter long. Use unit cubes to measure each object. Write the length of each object on the line.

Guided Practice

Use a centimeter ruler to measure in **centimeters**. Use a meterstick to measure in **meters**. There are 100 centimeters in a meter.

Helpful Hint
Use centimeters to measure shorter objects. Use meters to measure longer objects.

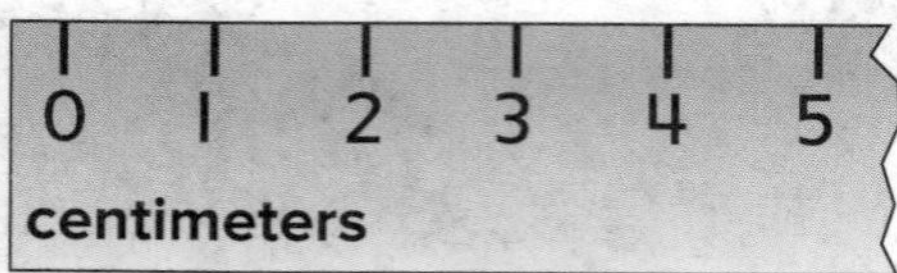

0 1 2 3 4 5
centimeters

A paper clip is about 5 centimeters long.

Find the object. Estimate the length. Measure each object in centimeters or meters.

Object	Estimate	Measure
1.	about ______ centimeters	about ______ centimeters
2.	about ______ meters	about ______ meters
3.	about ______ centimeters	about ______ centimeters

Talk Math Identify objects in the classroom that are about 1 centimeter long.

Name ....

Independent Practice

Find the object. Estimate the length. Measure each object in centimeters or meters.

Object	Estimate	Measure
4.	about ______ centimeters	about ______ centimeters
5.	about ______ centimeters	about ______ centimeters
6.	about ______ centimeters	about ______ centimeters
7.	about ______ meters	about ______ meters
8.	about ______ meters	about ______ meters

Problem Solving

9. Tyler runs the 100 meter dash two times. How many meters does he run in all?

_______ meters

10. Jessa's ski poles are 65 centimeters tall. Her sister's ski poles are 80 centimeters tall. How much shorter are Jessa's ski poles?

_______ centimeters

Brain Builders

11. Molly's wagon is 600 centimeters long. How many meters long is Molly's wagon?

_______ meters

12. The length of Isaac's bed is 2 meters. How many centimeters long is Isaac's bed? Explain.

____

____

____

Name ....

My Homework

Lesson 7

Centimeters and Meters

Homework Helper

Need help? connectED.mcgraw-hill.com

The cube is 1 centimeter long. There are 100 centimeters in 1 meter.

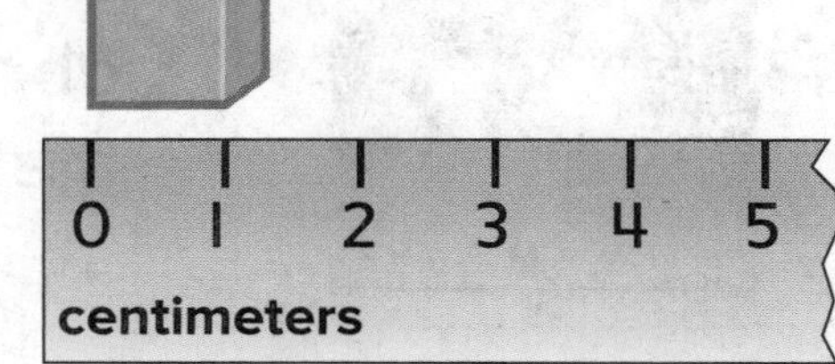

Practice

Find the object. Estimate the length. Measure each object in centimeters or meters.

Object	Estimate	Measure
1.	about ______ centimeters	about ______ centimeters
2.	about ______ centimeters	about ______ centimeters
3.	about ______ meters	about ______ meters

Find the object. Estimate the width. Measure each object in centimeters or meters.

Object	Estimate	Measure
4.	about ______ centimeters	about ______ centimeters
5.	about ______ meters	about ______ meters

Brain Builders

6. One pine tree is 4 meters tall. The other pine tree is 3 meters tall. How many centimeters taller is the first pine tree?

______ centimeters

Vocabulary Check

7. Circle the answer that is the same as 1 **meter**.

1 centimeter

10 centimeters

100 centimeters

1,000 centimeters

Math at Home Have your child identify objects he or she would measure using centimeters.

Name

Lesson 8
Select and Use Metric Tools

ESSENTIAL QUESTION
How can I measure objects?

Math in My World

Watch

1.

2.

Teacher Directions: Draw a picture of an object you would measure with a centimeter ruler in box 1. Draw a picture of an object you would measure with a meterstick in box 2.

Processes &Practices

Guided Practice

A centimeter ruler measures smaller objects. A meterstick measures larger objects. Circle the tool you would use to measure a lunchbox.

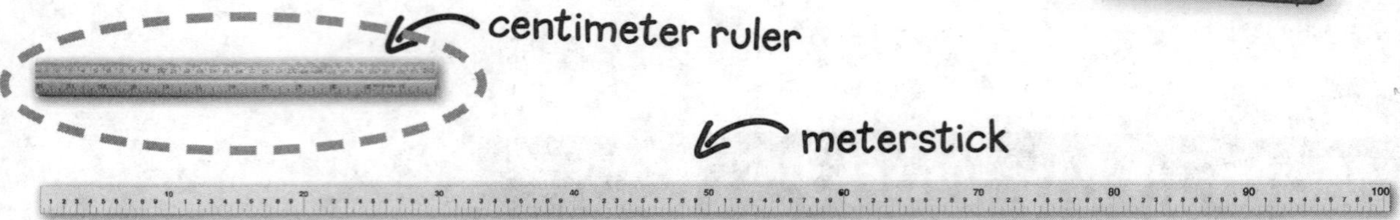

Find the object. Choose the tool and measure it. Explain why you chose that tool.

Object	Tool	Measure
1.	________	about ________
2.	________	about ________
3.	________	about ________

Talk Math Can you measure a paper clip with a meterstick? Explain.

Name ....

Independent Practice

Find the object. Choose the tool and measure it. Explain why you chose that tool.

Object	Tool	Measure
4.	________	about ________
5.	________	about ________
6.	________	about ________
7.	________	about ________
8.	________	about ________

Problem Solving

9. A wall in Jim's garage is 5 meters long. He is painting 2 meters of the wall blue. He wants to paint the rest of the wall red. How many meters will he paint red?

_______ meters

Brain Builders

10. A sandbox is 3 meters long. How many centimeters long is it?

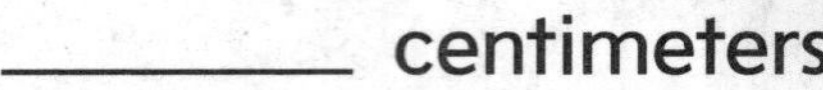

________ centimeters

11. Allison swam 50 meters in the morning. She swam 40 meters in the evening. How many meters did Allison swim in all? _______ meters

Write a number sentence that shows how you solved the problem.

____

Write Math Why do you need to know customary and metric length?

____

____

____

Name ________________________________

My Homework

Lesson 8

Select and Use Metric Tools

Homework Helper

Need help? connectED.mcgraw-hill.com

A centimeter ruler is used to measure smaller objects.

centimeter ruler

A meterstick is used to measure longer objects.

meterstick

Practice

Find the object. Choose the tool and measure it. Explain why you chose that tool.

Object	Tool	Measure
1.	____________	about ________
2.	____________	about ________

Find the object. Choose the tool and measure it. Explain why you chose that tool.

Object	Tool	Measure
3.	________	about________
4.	________	about________

Brain Builders

5. A tall slide at the pool is 10 meters high. A small slide is 4 meters high. How much higher is the tall slide?

________ meters

6. The hundred meter dash is marked with a blue line. The 400 meter dash is marked with a red line. How much farther is the red line?

________ meters

7. **Test Practice** Eight meters is equal to ________ centimeters.

400	600	800	900
○	○	○	○

Math at Home Have your child identify objects he or she would measure using meters and centimeters.

Name

Lesson 9

Compare Metric Lengths

ESSENTIAL QUESTION
How can I measure objects?

Explore and Explain

_____ centimeters

_____ centimeters

Teacher Directions: Find two objects in your classroom. Draw a picture of each object. Measure each object in centimeters. Write how many centimeters long each object is. Circle the object that is longer.

Processes &Practices

Guided Practice

You can compare lengths of objects.

76 centimeters

The blue tennis racket is ___69___ centimeters long.

The green tennis racket is ___76___ centimeters long.

The blue racket is ___7___ centimeters shorter than the green racket.

Find the objects. Measure them. Write the lengths. Write longer or shorter.

1.

_______ centimeters _______ centimeters

The lunchbox is _______ centimeters _______________.

2.

_______ meters _______ meters

The car is _______ meters _______________.

Name ....

Independent Practice

Find the objects. Measure them. Write the lengths. Write longer or shorter.

3. 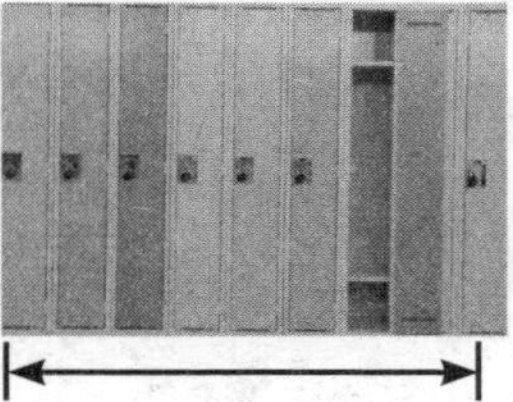

_______ meters _______ meters

The slide is _______ meters _____________.

4. 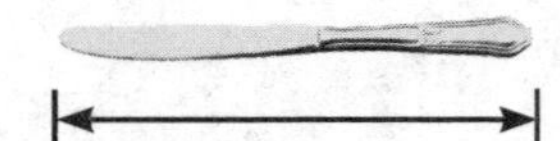

_______ centimeters _______ centimeters

The knife is _______ centimeters _____________.

5. 

_______ meters _______ meters

The swing set is _______ meters _____________.

6. 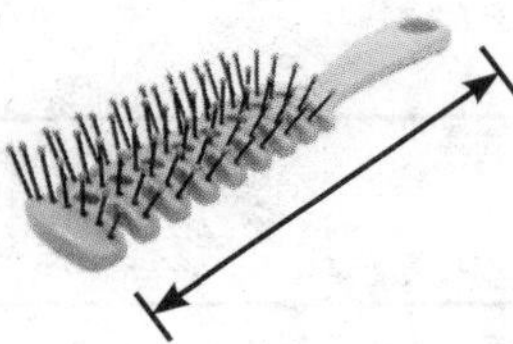

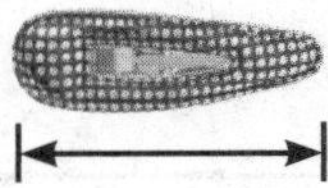

_______ centimeters _______ centimeters

The hair brush is _______ centimeters _____________.

Problem Solving

7. The red skateboard is 75 centimeters long. The purple skateboard is 83 centimeters long. How much longer is the purple skateboard?

_______ centimeters

8. Marcus ran 50 meters. Will ran 67 meters. How much farther did Will run?

_______ meters

Brain Builders

9. Alec rode his bike 82 meters. Kyla rode her bike 14 meters farther than Alec. How far did Kyla ride her bike? Explain your answer to a friend.

_______ meters

Write Math How can you compare a measurement in centimeters to a measurement in meters?

____

____

____

Name ________________________________

My Homework

Lesson 9

Compare Metric Lengths

Homework Helper

Need help? connectED.mcgraw-hill.com

You can compare lengths of objects.

The brown and white guinea pig is 3 centimeters longer.

Practice

Find the objects. Measure them. Write the lengths. Write longer or shorter.

1.

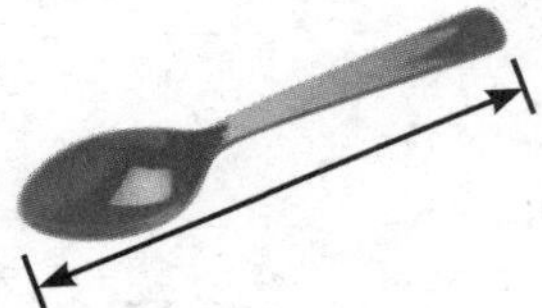

_______ centimeters _______ centimeters

The spoon is _______ centimeters _______________.

2.

_______ meters _______ meters

The house is _______ meters _______________.

Find the objects. Measure them. Write the lengths. Write longer or shorter.

3. 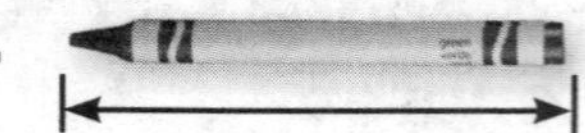

_______ centimeters _______ centimeters

The crayon is _______ centimeters _____________.

4. 

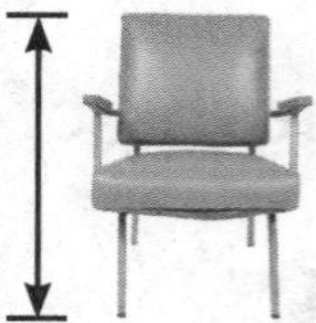

_______ meters _______ meters

The couch is _______ meters _____________.

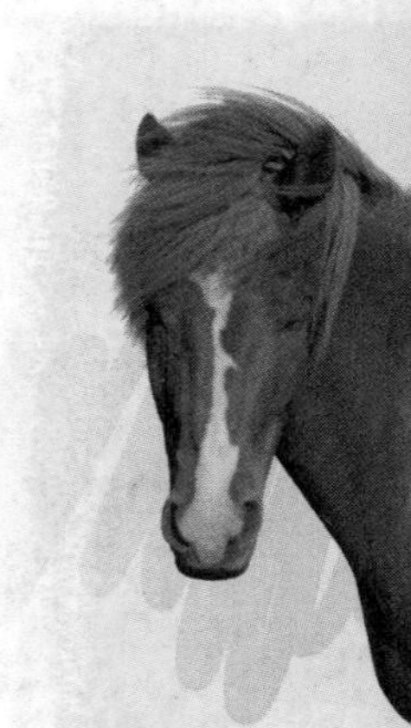

Brain Builders

5. The path from my house to the barn is 50 meters. The path from the barn to the pond is 30 meters. How much longer is the path to the barn? _______ meters

Write a number sentence that shows how you know.

____

6. **Test Practice** 56 meters + 35 meters = _____ meters.

86	87	91	95
○	○	○	○

Math at Home Have your child measure two beds in your house and compare the lengths in meters.

Name ....

Lesson 10

Relate Centimeters and Meters

ESSENTIAL QUESTION
How can I measure objects?

Math in My World

______ centimeters or ______ meters

Teacher Directions: Choose a large object from the picture such as a table or window. Find it in your school. Measure the object in centimeters. Measure the object in meters. Discuss the measurements.

Guided Practice

You can use different units of length to measure the same object.

The basketball hoop is about

___3___ meters tall.

It is about ___300___ centimeters tall.

Helpful Hint
1 meter = 100 centimeters

Find the object. Measure the length of each object twice.

Object	Measure
1.	______ centimeters ______ meters
2.	______ centimeters ______ meters
3.	______ centimeters ______ meters

Talk Math Which unit of measure gives you a more exact measurement?

Name

Independent Practice

Find the object. Measure the length of each object twice.

Object	Measure
4.	_______ centimeters _______ meters
5.	_______ centimeters _______ meters
6.	_______ centimeters _______ meters
7.	_______ centimeters _______ meters
8.	_______ centimeters _______ meters

Problem Solving

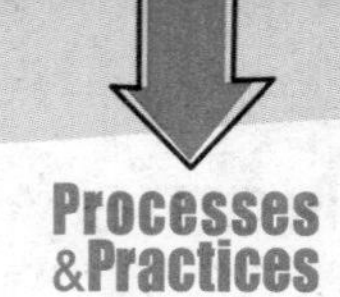

9. Kate's garden is 9 meters long. Stella's garden is 954 centimeters long. Whose garden is longer?

____

10. Brianna's bedroom is 6 meters wide. Oliver's bedroom is 596 centimeters wide. Whose bedroom is wider?

____

Brain Builders

11. Gianna jumped 1 meter. Parker jumped 75 centimeters. How many centimeters longer did Gianna jump than Parker?

________________ centimeters

12. Abby's ribbon was 5 meters long. She cut off 35 centimeters. How long is the piece of ribbon now? Explain.

____

____

____

____

Name ....

My Homework

Lesson 10

Relate Centimeters and Meters

Homework Helper

Need help? connectED.mcgraw-hill.com

You can use different units of length to measure the same object.

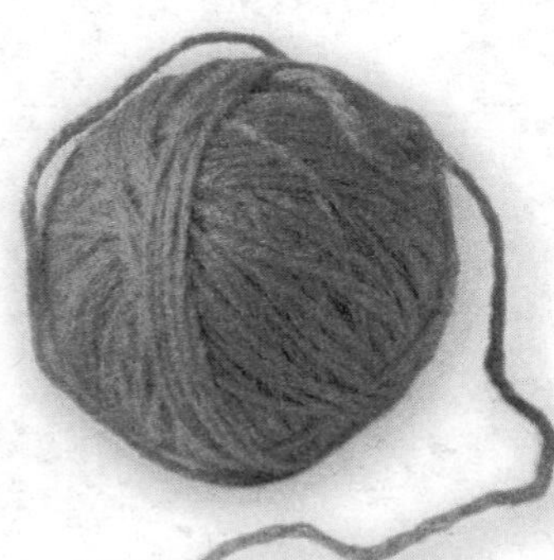

The yarn is 2 meters long.
It is 200 centimeters long.

Practice

Find the object. Measure the length of each object twice.

Object	Measure
1.	_______ centimeters _______ meters
2.	_______ centimeters _______ meters
3.	_______ centimeters _______ meters

Find the object. Measure the length of each object twice.

Object	Measure
4.	_______ centimeters _______ meters
5.	_______ centimeters _______ meters

Brain Builders

6. At the end of the race, Carly was 3 meters ahead of Reagan. How many centimeters was she ahead of Reagan?

_______ centimeters

7. **Test Practice** 35 centimeters + 65 centimeters = _______

I centimeter 

I meter 

90 centimeters 

2 meters 

Math at Home Have your child measure an object in your house using centimeters and meters. Discuss how each measurement relates to the size of the unit.

Name ....

Lesson 11
Measure on a Number Line

ESSENTIAL QUESTION
How can I measure objects?

Math in My World

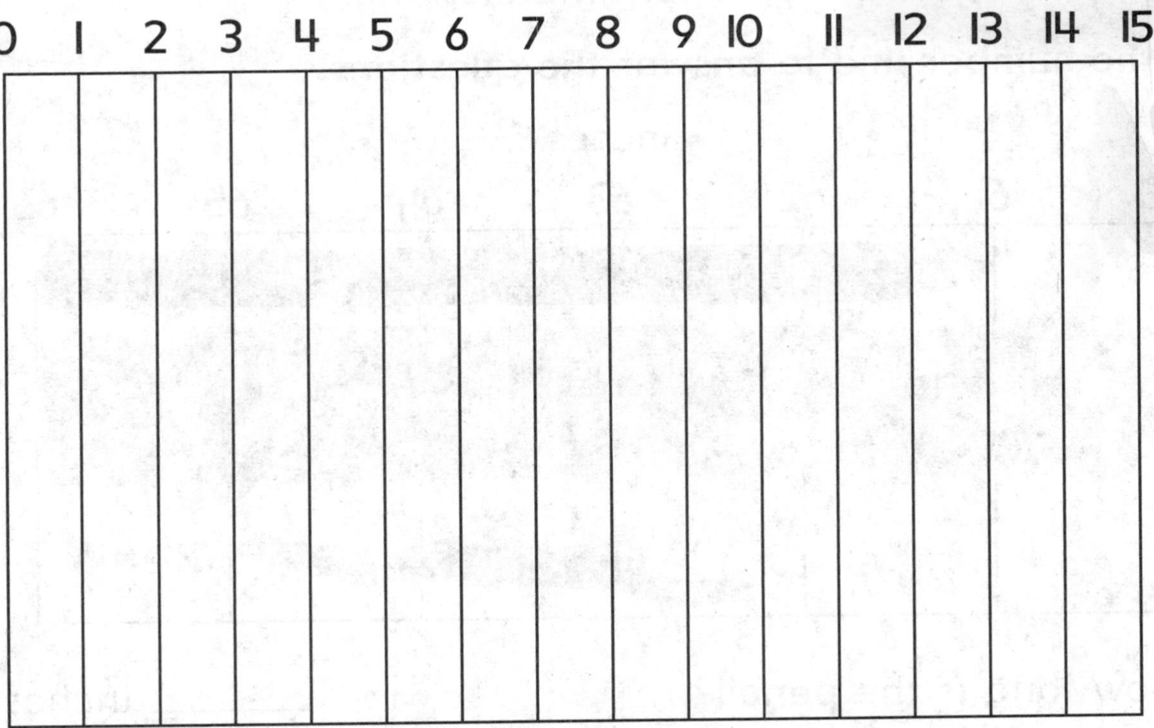

_____ – _____ = _____

Teacher Directions: Place a crayon and a pair of scissors on the number line. Line up the ends at 0. How much longer are the scissors than the crayon? Repeat with other pairs of objects. Draw a pair of objects on the number line and write the subtraction sentence that shows how much longer one object is than another.

Processes &Practices

Guided Practice

centimeters

20 21 22 23 24 25 26 27 28 29 30

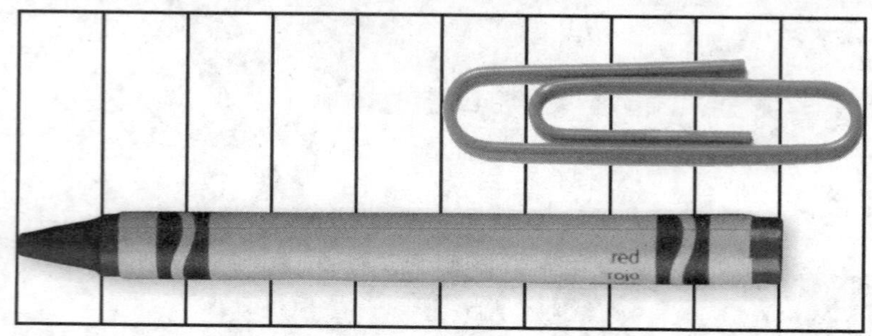

The crayon is ___9___ centimeters long.

The paper clip is ___5___ centimeters long.

Use the number line to answer the questions.

inches

60 61 62 63 64 65 66

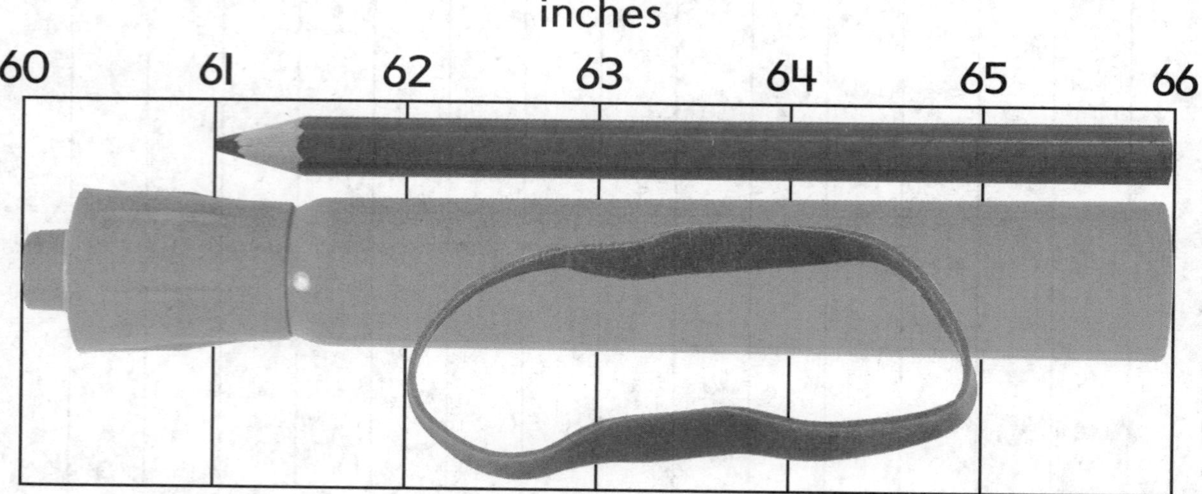

1. How long is the pencil? _______ inches

2. How much longer is the marker than the rubber band? _______ inches

Talk Math How does the number line help you compare measurements?

Name

Independent Practice

Use the number line to answer the questions.

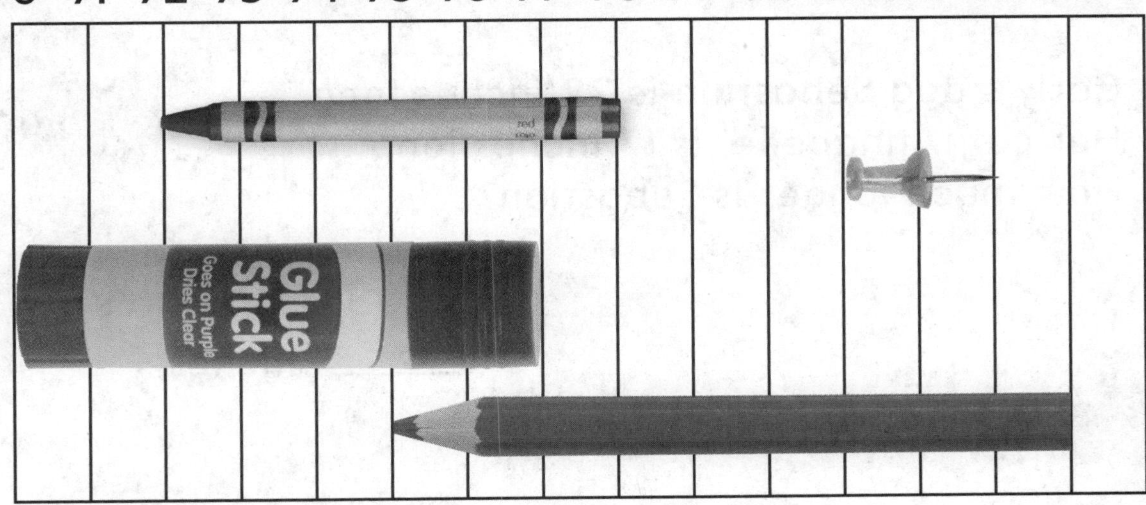

3. How long is the glue stick? _______ centimeters

4. How long is the push pin? _______ centimeters

5. How long is the crayon? _______ centimeters

6. How much longer is the pencil than the glue stick? _______ centimeters

7. How long are the crayon and push pin in all? _______ centimeters

8. How much shorter is the push pin than the pencil? _______ centimeters

Problem Solving

9. Makenna is 38 inches tall. Her brother is 26 inches tall. How much taller is Makenna?

 inches

10. Carly's dog Sebastian is 22 inches long. Her cat, Annabelle, is 17 inches long. How much longer is Sebastian?

 inches

Brain Builders

11. Levi found a trail in the woods that is 54 meters long. Then he found a second trail that is 29 meters longer than the first trail. How long is the second trail?

_______ meters

Write Math How can a number line help you measure objects? Explain.

____

____

____

Name ....

Lesson 11

Measure on a Number Line

My Homework

Homework Helper 

Need help? connectED.mcgraw-hill.com

inches

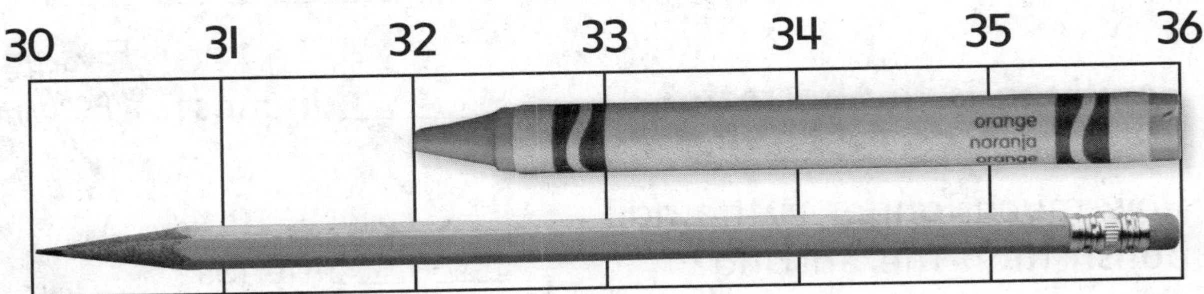

The pencil is 2 inches longer than the crayon.

Practice

Use the number line to answer the questions.

centimeters

60 61 62 63 64 65 66 67 68 69 70 71 72 73 74 75

1. How long is the truck? ______ centimeters

2. How long is the car? ______ centimeters

Use the number line to answer the questions.

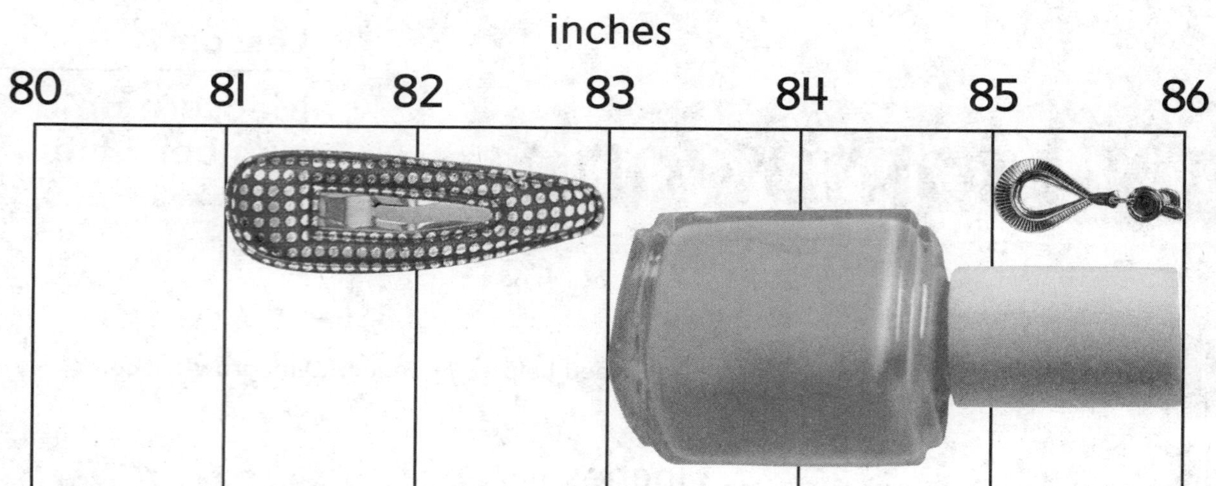

Favorite color: Peppermint Stick!

3. How long is the barrette? _______ inches

4. How much longer is the nail polish than the earring? _______ inches

Brain Builders

5. Sean builds a block tower that is 32 inches high. Victoria builds a block tower that is 3 feet high. Whose tower is higher? __________

How much higher is it? _______ inches

6. **Test Practice** 36 inches + 26 inches = _______

62 inches	62 centimeters	62 feet	62 meters
○	○	○	○

Math at Home Help your child create a large number line measured in inches. Have him or her measure objects from your home using the number line.

Name

Lesson 12
Measurement Data

ESSENTIAL QUESTION
How can I measure objects?

Math in My World

Pencil Lengths

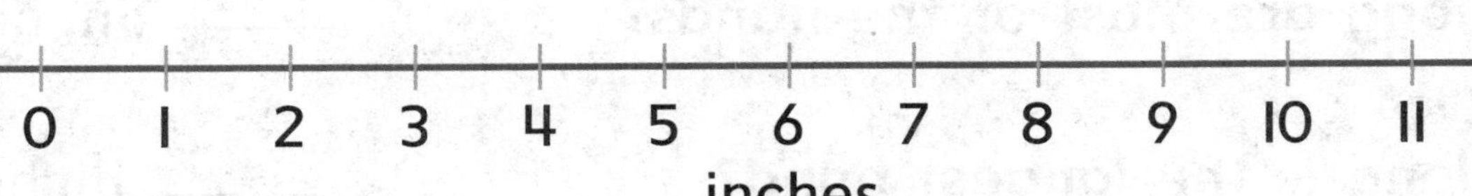

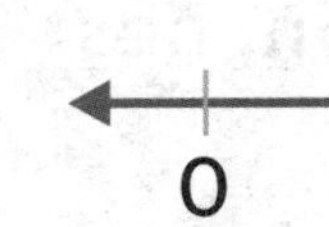

Teacher Directions: Measure five of your classmates' pencils with a ruler. Make a line plot using the data from your measurements.

Processes & Practices

Guided Practice

Helpful Hint
Remember a line plot tells how many times a number occurs in data.

Shoe Lengths

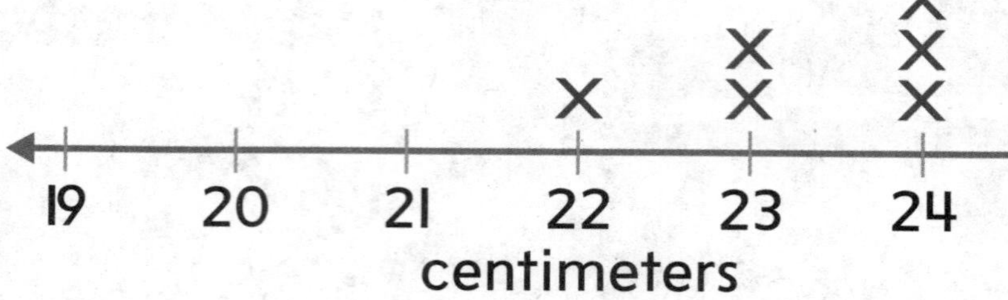

How many people have shoes that are 23 centimeters long? 2

Measure the length of the right hand of 10 people in inches. Use the data to make a line plot.

Hand Lengths

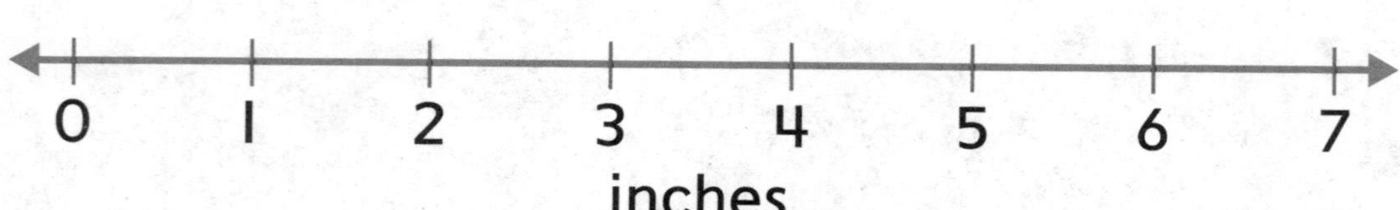

1. **How long are most of the hands?** ______ inches

2. **How long is the longest hand?** ______ inches

Talk Math How does a line plot help to show measurement data?

Name

Independent Practice

Measure 15 books from your classroom library in centimeters. Use the data to make a line plot.

Book Lengths

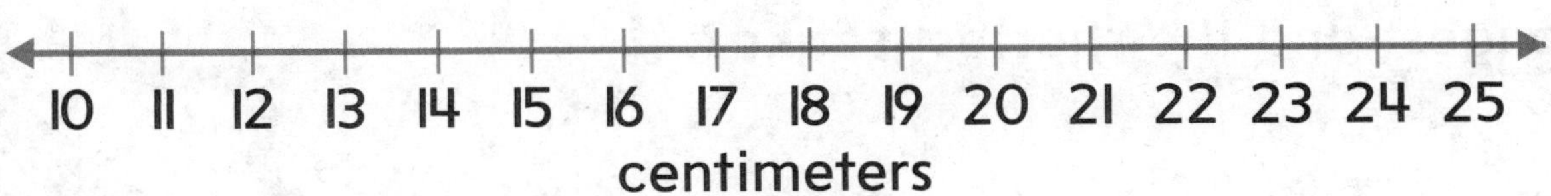

3. How long are most books? ______ centimeters

4. How long is the longest book? ______ centimeters

Measure 15 used crayons in inches. Use the data to make a line plot.

Crayon Lengths

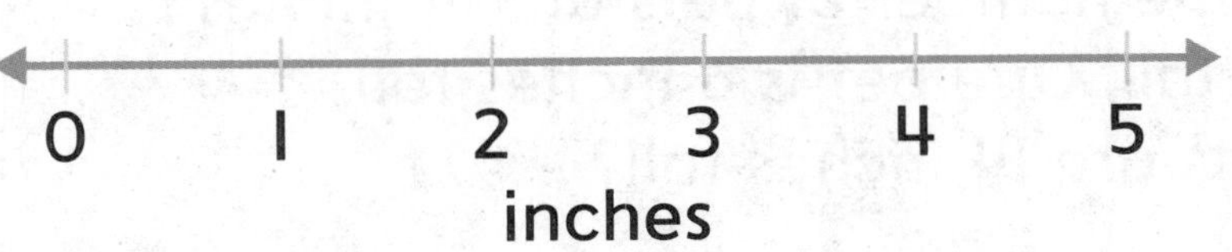

5. How short is the shortest crayon? ______ inches

6. How long is the longest crayon? ______ inches

Problem Solving

7. A line plot shows that 12 people are 48 inches tall, 8 people are 50 inches tall, and 9 people are 47 inches tall. How tall are most of the people?

_______ inches

8. Four snakes are 20 inches, 2 snakes are 32 inches, and 4 snakes are 15 inches. What is the difference in length between the longest and the shortest snake?

_______ inches

Brain Builders

9. Four people went down the slide 8 times. Eight people went down the slide 6 times. Two people went down the slide 12 times. How many times did most people go down the slide? Explain your answer to a friend.

_______ times

10. Jordan made a line plot of the heights of his 5 pets. The same number of pets are 14 inches and 18 inches tall. One pet is 6 inches tall. How many pets are 14 inches tall?

____

____

____

Name

Lesson 5

Regroup Hundreds

My Homework

Homework Helper

Need help? connectED.mcgraw-hill.com

Find 436 − 245.

Step 1 Subtract ones.

Step 2 Subtract the tens. You cannot subtract 4 from 3. Regroup 1 hundred as 10 tens.

Step 3 Subtract the hundreds.

	hundreds	tens	ones
	3	13	
	~~4~~	~~3~~	6
−	2	4	5
	1	9	1

Practice

Subtract.

1.

	hundreds	tens	ones
	□	□	
	3	6	2
−	2	7	1

2.

	hundreds	tens	ones
	□	□	
	7	3	2
−	4	4	1

3.

	□	□	
	8	3	5
−		5	5

4.

	□	□	
	7	3	5
−	6	7	2

5.

	□	□	
	2	1	7
−	1	4	4

Name

My Homework

Homework Helper

Need help? connectED.mcgraw-hill.com

Three people have pencils that are 8 centimeters long.

Two people have pencils that are 13 centimeters long.

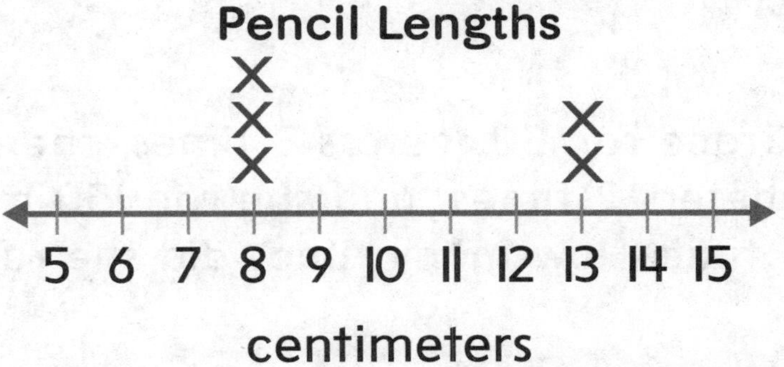

Practice

Measure the thumb length of each of your family members. Use the data to make a line plot.

Helpful Hint
Line plots show how often a number occurs in data.

Thumb Lengths

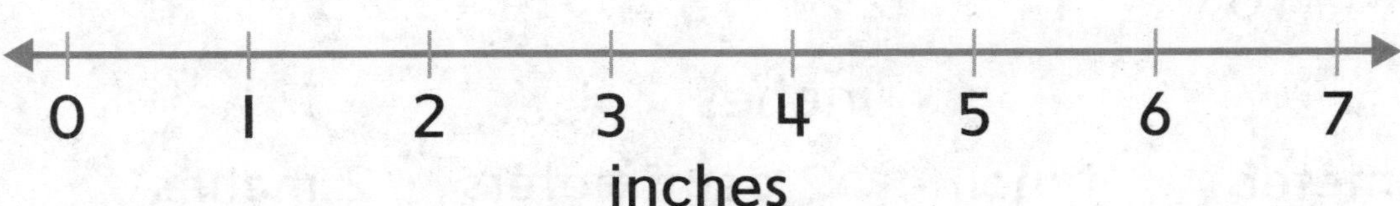

1. How long is the longest thumb? ______ inches

2. How short is the shortest thumb? ______ inches

3. Make a line plot to show how tall each flower is. Four flowers are 10 centimeters, 3 flowers are 13 centimeters, and 6 flowers are 17 centimeters.

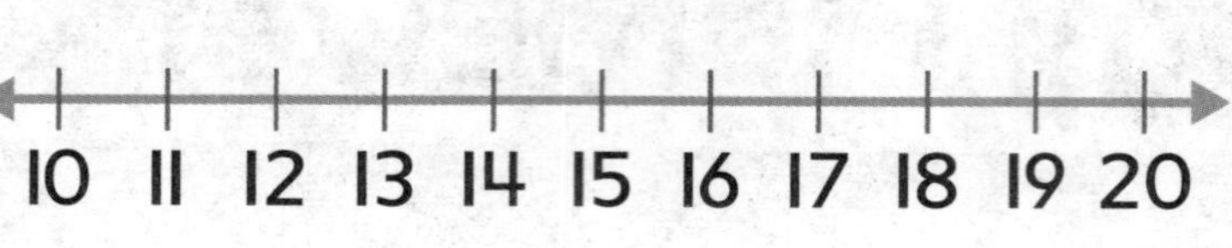

Brain Builders

4. Grace ran 50 meters 3 times, she ran 100 meters 2 times, and she ran 150 meters 1 time. How many times did she run in all?

______ times

5. **Test Practice** Use the line plot below to answer the question. How long are most toes?

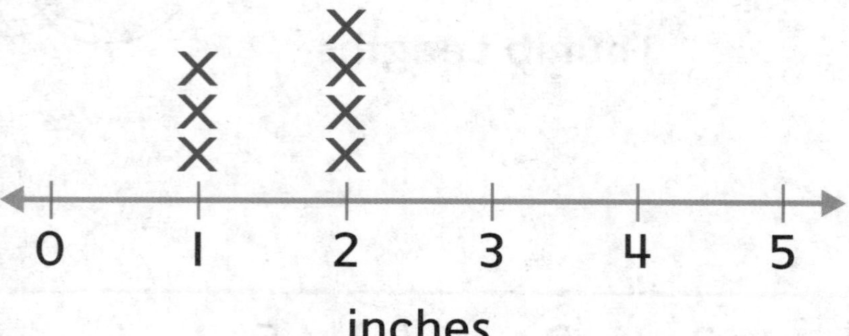

1 centimeter	1 inch	2 centimeters	2 inches
○	○	○	○

Math at Home Have your child create a line plot to show the heights of everyone in your family.

Name ....

My Review

Chapter 11

Customary and Metric Lengths

Vocabulary Check

Write the correct word in each blank.

length	inch	estimate	measure
foot	yard	centimeter	meter

1. You can __________ to find a number close to an exact amount.

2. __________ is how long or how far away something is.

3. A paper clip is about one __________ long.

4. A __________ is equal to 3 feet.

5. To __________ is to find the length, height, or weight using standard or nonstandard units.

6. A base-ten cube is about one __________ long.

7. A __________ is equal to 12 inches.

8. A __________ is equal to 100 centimeters.

Concept Check

Find the object. Estimate the length. Measure the object in centimeters.

Object	Estimate	Measure
9.	about ______ centimeters	about ______ centimeters

Find the object. Choose the tool and measure it.

Object	Tool	Measure
10.	______	about ______

Find the objects. Measure them. Write the lengths. Write longer or shorter.

11.

______ centimeters ______ centimeters

The glue stick is ______ centimeters ______.

Find the object. Measure the length of it twice.

Object	Measure
12.	______ centimeters ______ meters

Name

Problem Solving

13. Use the data to make a line plot.
6 pieces of yarn were 4 inches long,
3 pieces of yarn were 2 inches long,
2 pieces of yarn were 6 inches long,
and 1 piece of yarn was 5 inches long.

Yarn Lengths

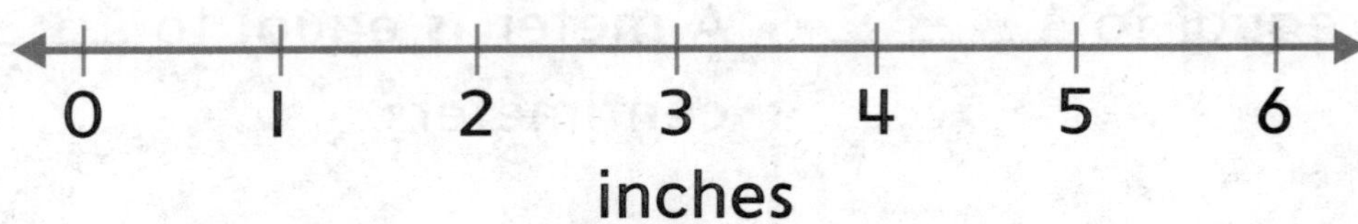

Brain Builders

14. Mason's bike is 1 yard long. His scooter is 2 feet long. How many inches longer is his bike than his scooter?

_______ inches

15. **Test Practice** Lucy's lacrosse stick is 1 meter long. How many centimeters long is Lucy's lacrosse stick?

50	100	200	400
○	○	○	○

Reflect

Chapter 11

Answering the Essential Question

Completed each sentence.

ESSENTIAL QUESTION How can I measure objects?

Customary	Metric
A foot is equal to _______ inches.	A meter is equal to _______ centimeters.
A yard is equal to _______ inches or _______ feet.	My foot is about _______ centimeters long.
My foot is about _______ inches long.	I can measure metric lengths with a _______ or a _______.
I can measure customary lengths with a _______, _______, or _______.	

We are winners!

Name

Date

Score

Performance Task

Inch by Inch

The following table shows the length in feet of certain musical instruments.

Instrument	Length in Feet
Violin	1
Flute	2
Guitar	3
String Bass	6

Show all your work to receive full credit.

Part A

Rewrite the length of the guitar and the string bass in inches and yards. What do you notice about the relationship between unit size and the number of the unit that makes the length?

Part B

Which two instruments have the greatest difference in size? Which two instruments have the least difference in size? Explain.

Part C

Suppose that a musical group has 4 violins, 1 flute, 2 guitars, and 1 string bass with the measurements shown in the table. Make a line plot to show the lengths in feet of these instruments.

Chapter

Geometric Shapes and Equal Shares

ESSENTIAL QUESTION

How do I use shapes and equal parts?

Watch a video!

Name

Chapter 12 Project

Two-Dimensional Shape Collages

1. You will create a two-dimensional shape collage as you work through the chapter.
2. You may use two-dimensional shapes cut out of paper or magazines. Or you may also draw your own two-dimensional shapes.
3. For your collage, label the sides and angles of at least one of the two-dimensional shapes.
4. Your teacher will display your collage along with the collages from the other students in your classroom on a bulletin board to use as reference tools throughout the chapter.
5. Use the collages on the bulletin board to fill in the table below.

Shape	Number of Sides	Number of Angles
circle		
hexagon		
square		
rectangle		
triangle		
pentagon		
quadrilateral		
parallelogram		
trapezoid		

Name

Am I Ready?

Draw an X over the object that is a different shape.

1.   

2.   

Draw a line to match the objects that are the same shape.

3.  

4.  

5.  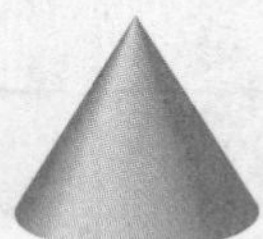

6. What shape is the game board?
Circle the word.

triangle square circle

How Did I Do?

Shade the boxes to show the problems you answered correctly.

1	2	3	4	5	6

Name

My Math Words

Review Vocabulary

circle rectangle square

Use the review vocabulary to complete the chart. The first row is done for you.

Word	Example	Not an Example
triangle		
____________		
____________		
____________		

My Vocabulary Cards

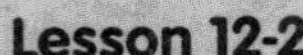

Lesson 12-2

angle

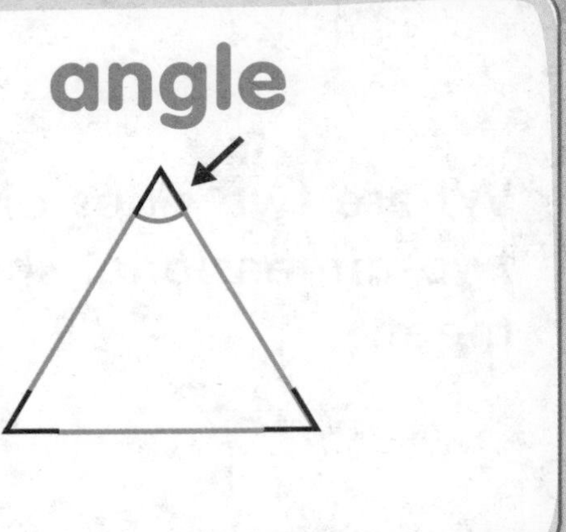

Lesson 12-4

cone

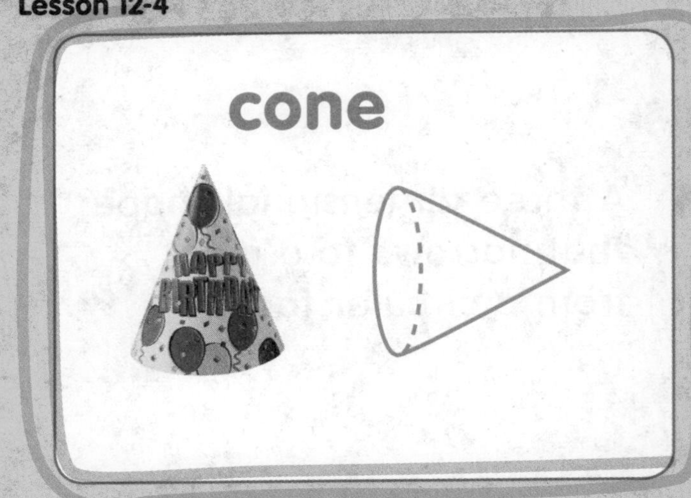

Lesson 12–4

cube

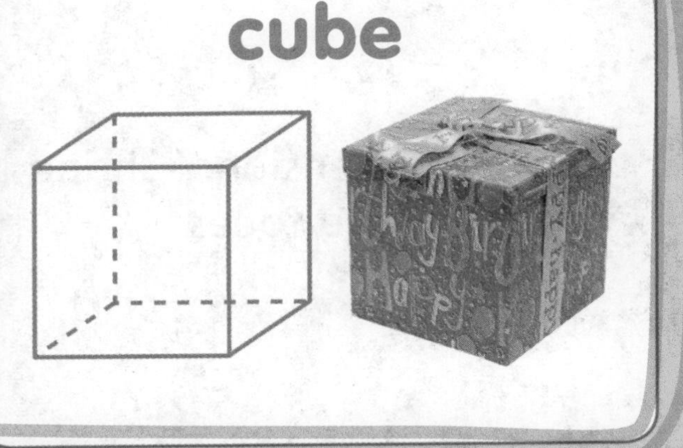

Lesson 12-4

cylinder

Lesson 12-5

edge

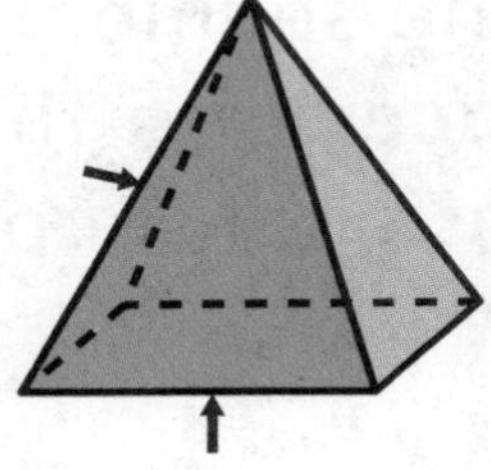

Lesson 12-5

face

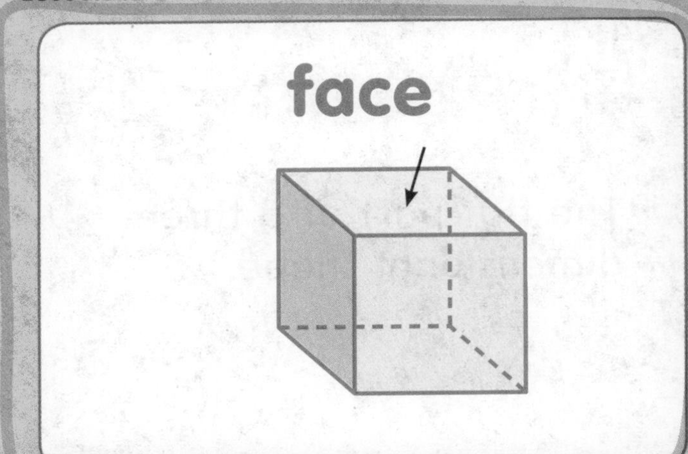

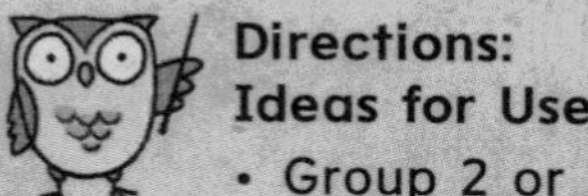

Directions:
Ideas for Use

- Group 2 or 3 common words. Add a word that is unrelated to the group. Ask another student to name the unrelated word.
- Have students draw an example for each card. Have them make drawings that are different from what is shown on each card.

A three-dimensional shape that narrows to a point from a circular face.

Where two sides on a two-dimensional shape meet.

A three-dimensional shape that is shaped like a can.

A three-dimensional shape with 6 square faces.

The flat part of a three-dimensional shape.

The line segment where two faces of a three-dimensional shape meet.

My Vocabulary Cards

Lesson 12-7

fourths

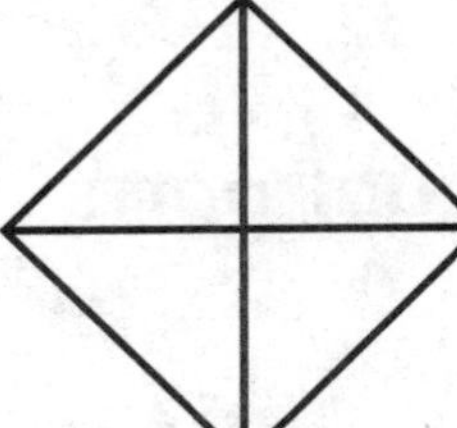

4 fourths

Lesson 12-7

halves

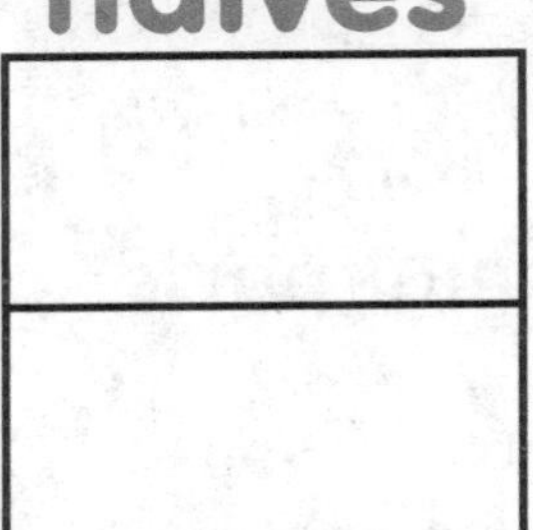

2 halves

Lesson 12-1

hexagon

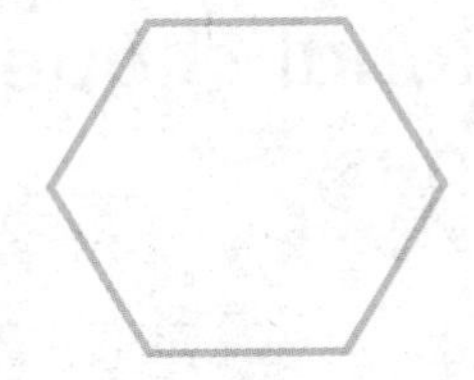

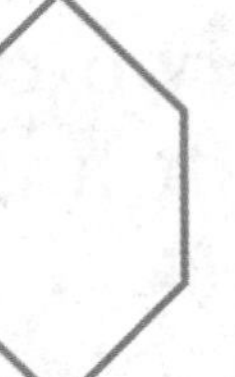

Lesson 12-1

parallelogram

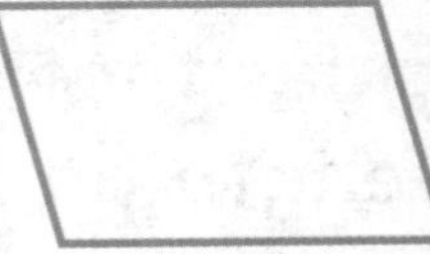

Lesson 12-7

partition

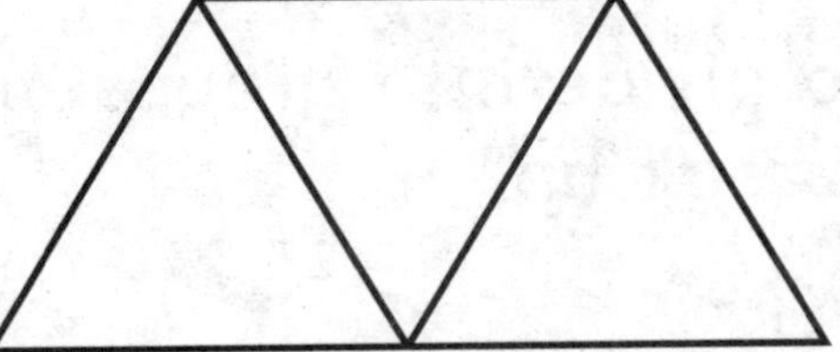

Lesson 12-2

pentagon

Directions:
More Ideas for Use

- Create a riddle for each word. Have a friend guess the word for each riddle.
- Have students write a tally mark on each card every time they read the word in this chapter or use it in their writing. Challenge them to try to use at least 10 tally marks for each word card.

Two equal parts

Four equal parts

A two-dimensional shape that has two pairs of sides that are the same length and are equal distance apart.

A two-dimensional shape that has six sides.

A polygon with five sides.

To divide into groups or "break up."

My Vocabulary Cards

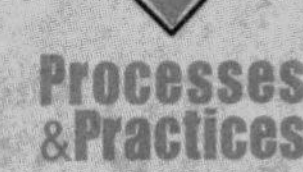

Lesson 12-4

pyramid

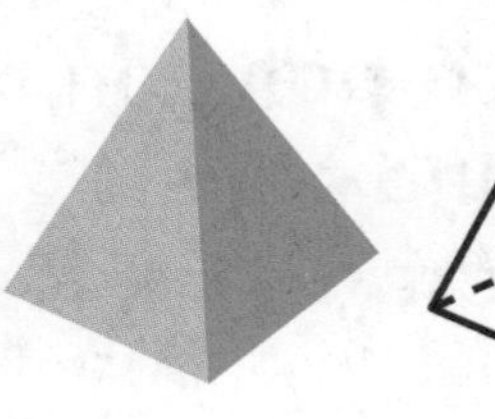

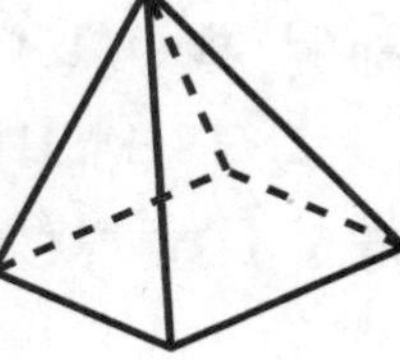

Lesson 12-1

quadrilateral

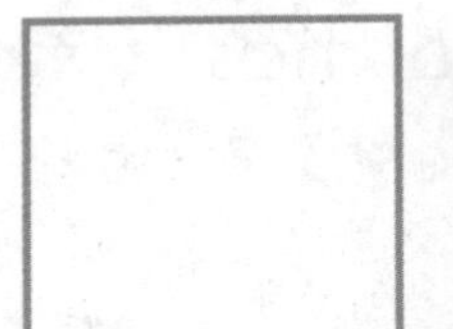

Lesson 12-2

rectangular prism

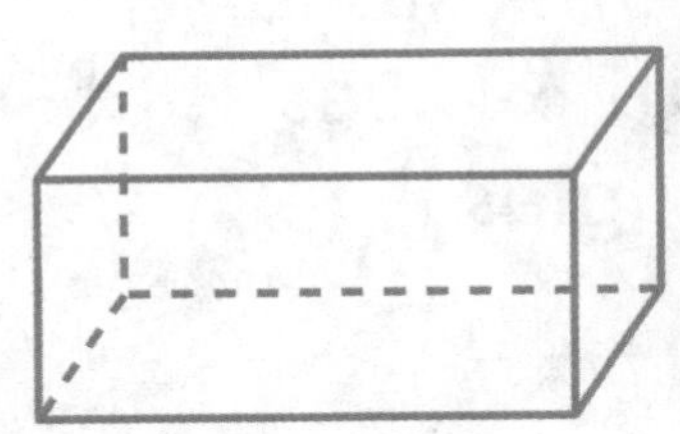

Lesson 12-4

side

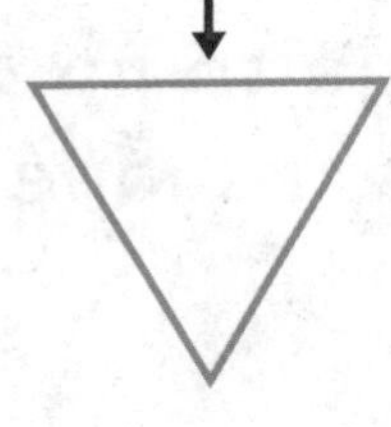

Lesson 12-6

sphere

Lesson 12-1

thirds

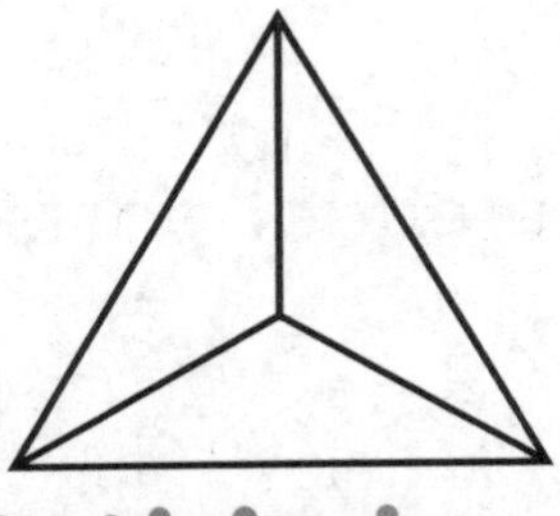

3 thirds

Directions:
More Ideas for Use

- Have students sort the words by the number of syllables in each word.
- Ask students to find pictures to show each word. Have them work with a friend to guess which word the picture shows.

A shape that has 4 sides and 4 angles.

A three-dimensional shape with a polygon as a base and other faces that are triangles.

One of the line segments that make up a shape.

A three-dimensional shape with 6 faces that are rectangles.

Three equal parts.

A three-dimensional shape that has the shape of a round ball.

My Vocabulary Cards

Lesson 12-4

three-dimensional shapes

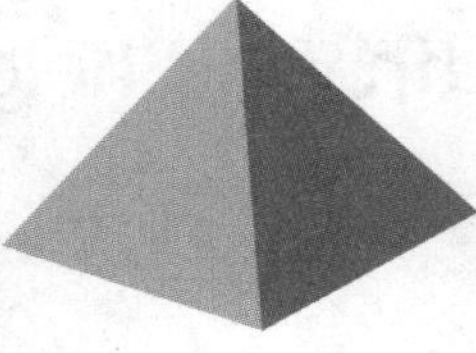

Lesson 12-7

trapezoid

Lesson 12-1

two-dimensional shapes

Lesson 12-1

vertex

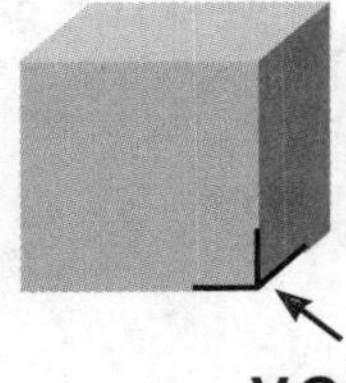

Directions:

More Ideas for Use

- Have students arrange the cards in alphabetical order.
- Group 2 or 3 common words. Add a word that is unrelated to the group. Ask another student to name the unrelated word.

A two-dimensional shape with four sides and only two opposite sides that are parallel.

A shape that has: length, width, and height.

The point where three or more faces of a three-dimensional shape meet.

A shape having the two dimensions length and width.

My Foldable

FOLDABLES® Follow the steps on the back to make your Foldable.

FOLDABLES
Study Organizer

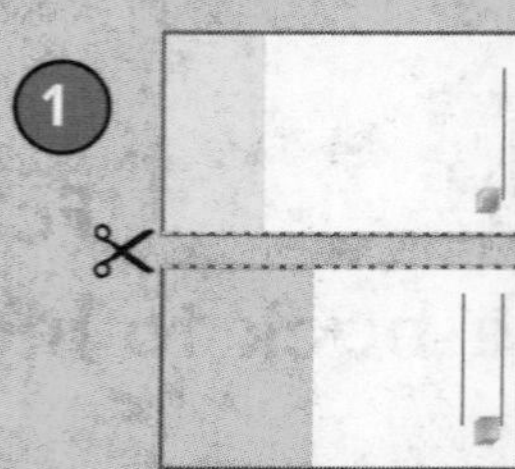

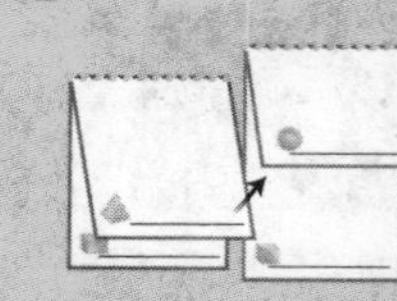

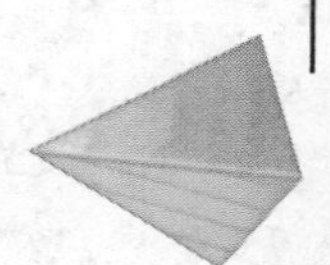

Name

Lesson 1
Two-Dimensional Shapes

ESSENTIAL QUESTION
How do I use shapes and equal parts?

Math in My World

circle

hexagon

square

rectangle

triangle

Teacher Directions: Use small attribute blocks. Trace and identify each shape. Draw a line from each shape to its name.

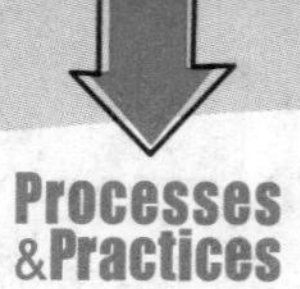

Guided Practice

A **two-dimensional shape** is a shape with only length and width.

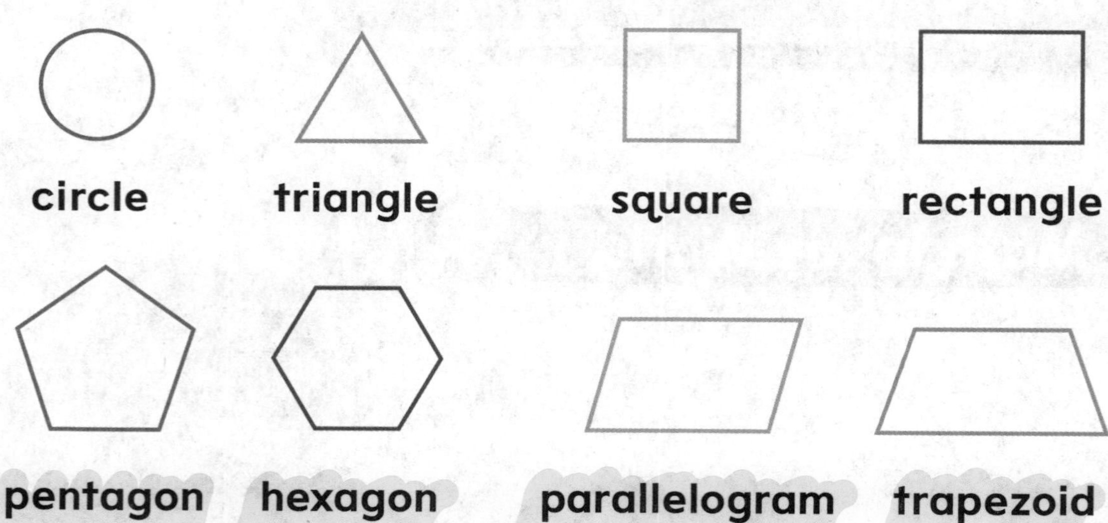

Circle the shapes that match the name.

1. parallelogram

2. triangle

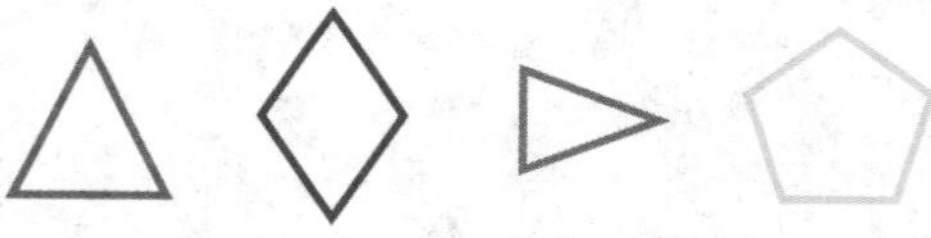

Write the name of the shape.
Circle the shape that matches.

3.

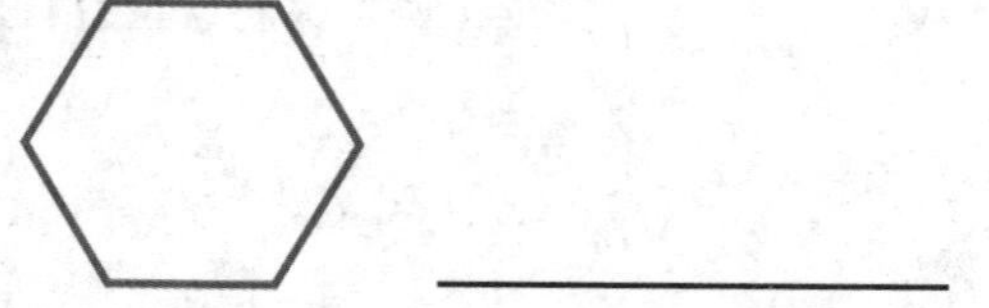

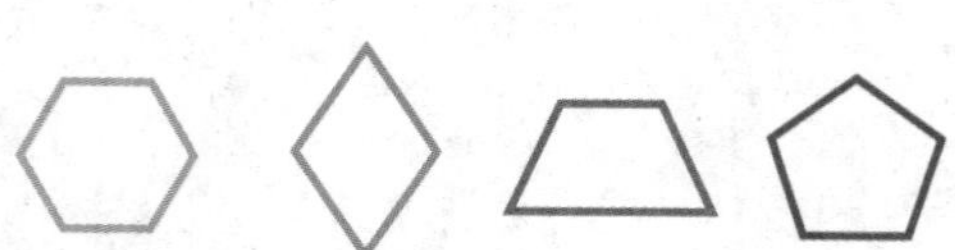

Talk Math What is the difference between a pentagon and a hexagon? How are they alike?

Name

Independent Practice

Circle the shapes that match the name.

4. trapezoid

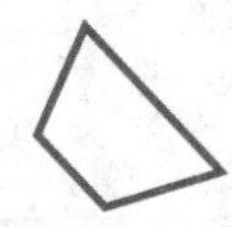  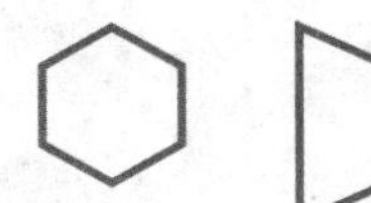

5. hexagon

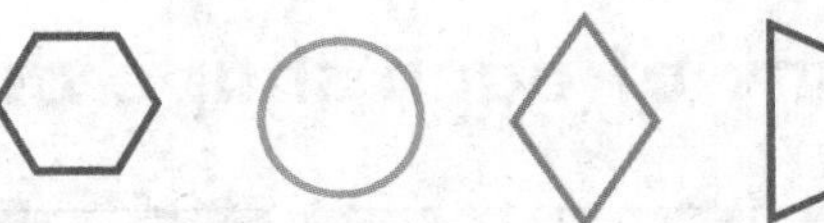

6. triangle

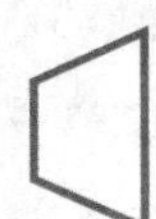  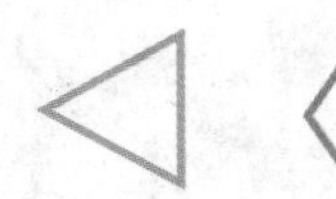 

7. pentagon

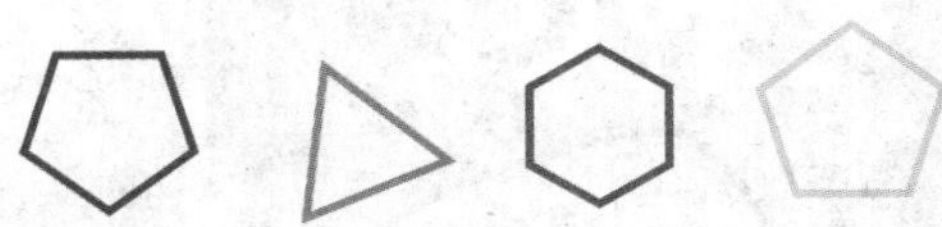

Write the name of the shape.
Circle the shape that matches.

8.

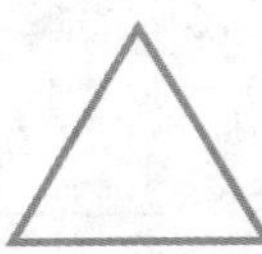 ____________

 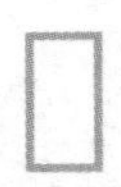  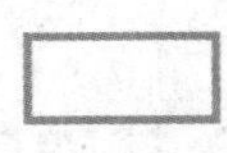

9.

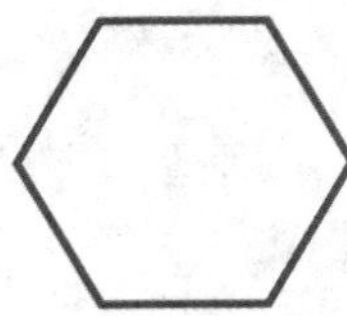 ____________

 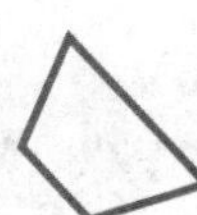  

Circle the shape that does not belong in each group.

10.   

11.  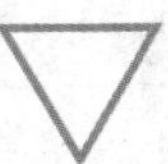  

Problem Solving

12. Identify the shape of each sign.

SPEED LIMIT 75

__________ __________

How many of each shape do you see?

13.

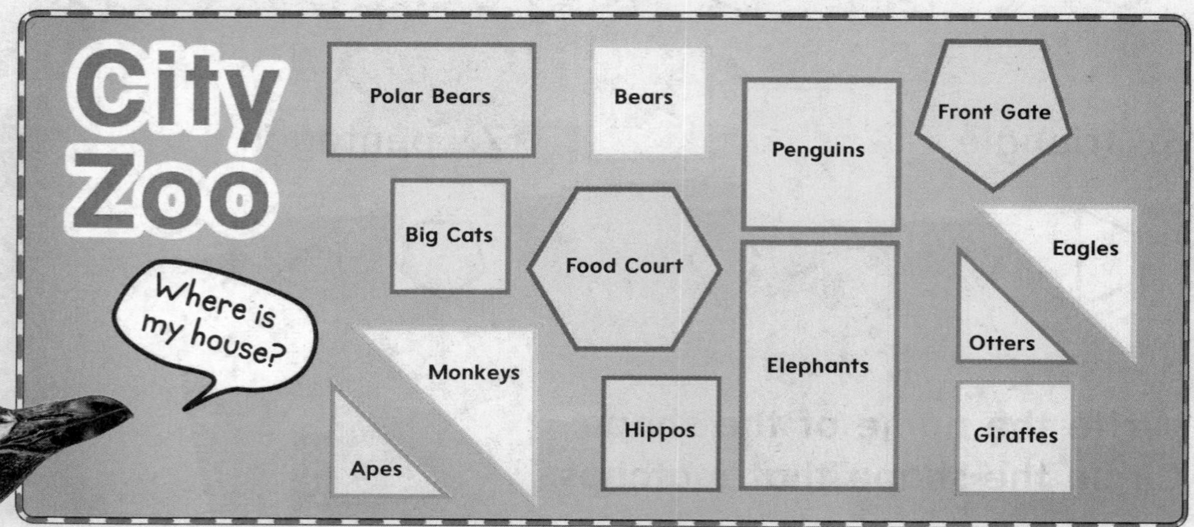

triangles ______ hexagons ______ rectangles ______

squares ______ pentagons ______ circles ______

Brain Builders

Write Math Give examples of objects in your school that look like triangles and squares.

____

____

____

Name ....

Lesson 1

Two-Dimensional Shapes

My Homework

Homework Helper

Need help? connectED.mcgraw-hill.com

A two-dimensional shape is a shape with only length and width.

circle

triangle

square

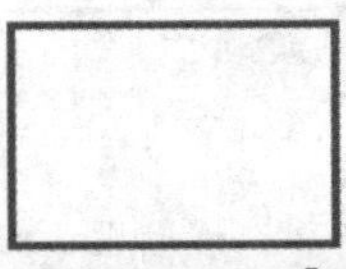

rectangle

pentagon

hexagon

parallelogram

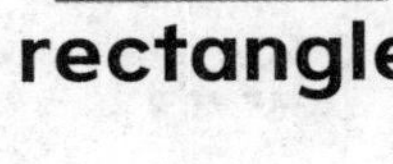

trapezoid

Practice

Circle the shapes that match the name.

1. rectangle

2. triangle

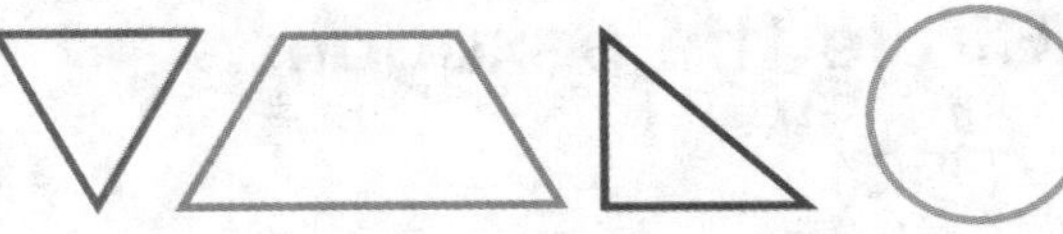

3. trapezoid

4. hexagon

Write the name of the shape. Circle the shape that matches.

5. 

____

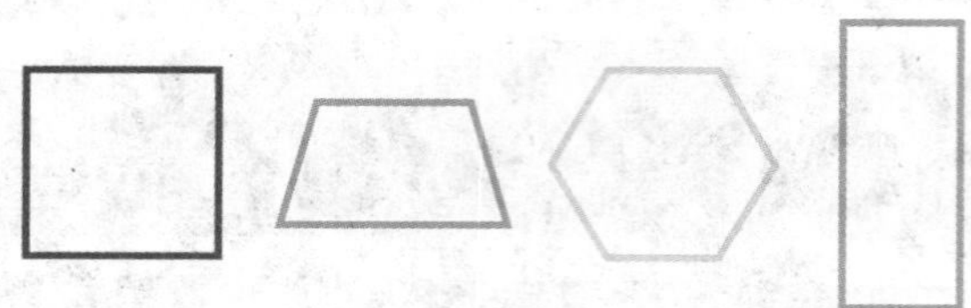

6. 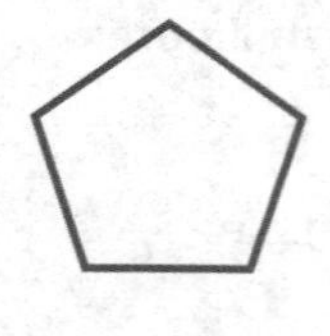

____

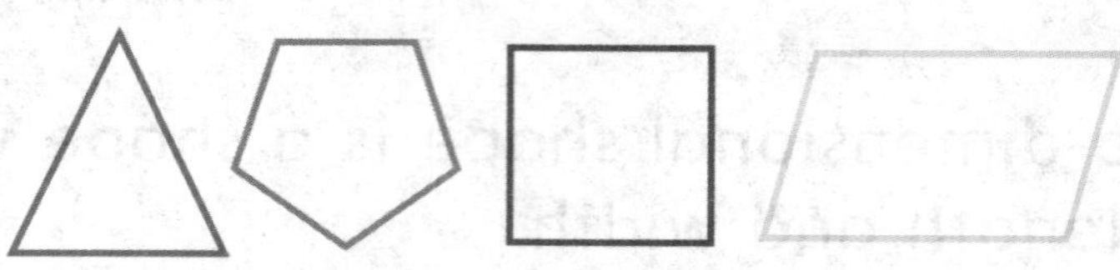

Brain Builders

mmm, blueberry!

7. Jack cut out a shape to glue onto a picture. The shape looked like an ice cream cone. Jack says he cut out a square. Tell why he is wrong.

____

____

Vocabulary Check 

8. Circle the hexagons.

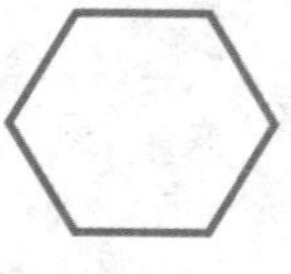
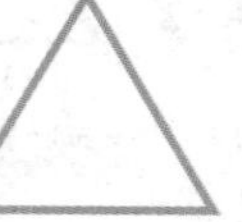

Math at Home Point to two-dimensional shapes around your house (triangles, squares, rectangles, hexagons, and pentagons) and have your child identify each shape.

Name

Lesson 2
Sides and Angles

ESSENTIAL QUESTION
How do I use shapes and equal parts?

Math in My World

_____ sides _____ angles

_____ sides _____ angles

_____ sides _____ angles

Teacher Directions: Have students sort triangle, square, parallelogram, trapezoid, and hexagon pattern blocks by their number of sides and angles. Trace them. Write how many sides and angles.

Processes & Practices

Guided Practice

You can describe two-dimensional shapes by the number of **sides** and **angles**.

triangle

side

angle

______ sides

______ angles

quadrilateral

______ sides

______ angles

pentagon

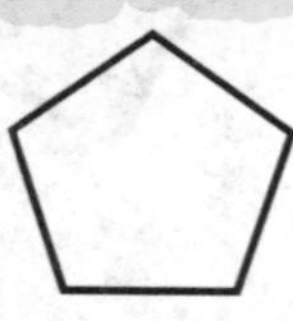

______ sides

______ angles

hexagon

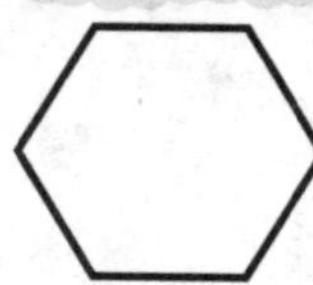

______ sides

______ angles

circle

______ sides

______ angles

Trace each shape. Write how many sides and angles.

1.

______ sides

______ angles

2.

______ sides

______ angles

3. Circle the objects that have 0 sides and 0 angles.

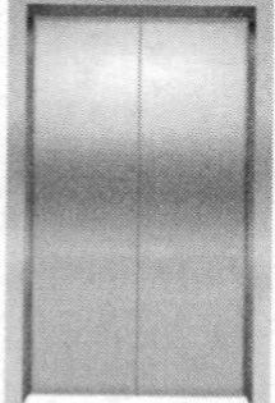

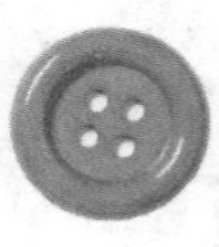

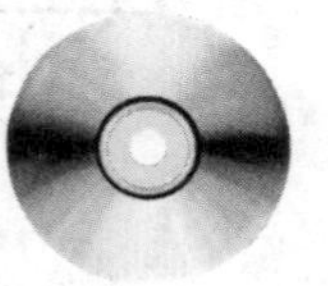

Talk Math How are a square and a hexagon alike? How are they different?

Name

Independent Practice

Trace each shape. Write how many sides and angles.

4. 

_____ sides

_____ angles

5. 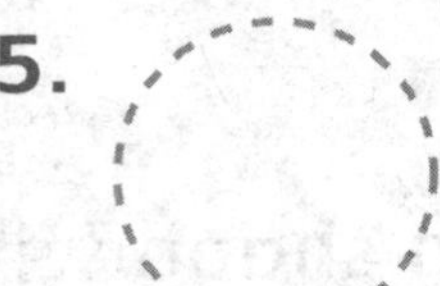

_____ sides

_____ angles

6. 

_____ sides

_____ angles

7. 

_____ sides

_____ angles

Circle the objects that match the description.

8. 3 sides and 3 angles

9. 4 sides and 4 angles

Problem Solving

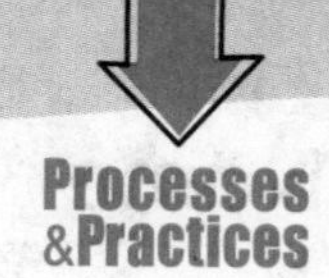

Draw a picture to solve.

10. Kira draws a shape with 6 sides and 6 angles. What shape does she draw?

____

11. Alex draws a shape with 3 sides and 3 angles. What shape does Alex draw?

____

Brain Builders

12. Josh drew 3 squares. Katie drew 2 triangles and 1 square. How many more angles did Josh draw than Katie?

_______________ angles

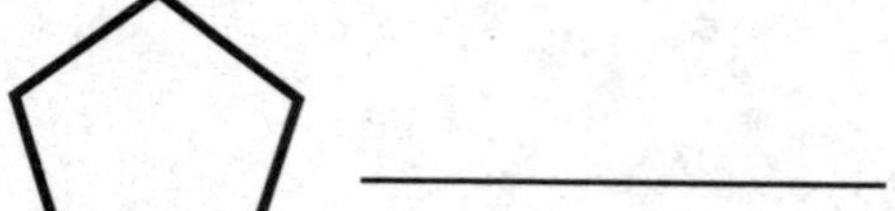

Write the name of each shape. Describe two things about each shape.

____

1. _______________________________

2. _______________________________

____

1. _______________________________

2. _______________________________

Name ....

My Homework

Lesson 2

Sides and Angles

Homework Helper

Need help? connectED.mcgraw-hill.com

A two-dimensional shape can be described by its sides and angles.

triangle

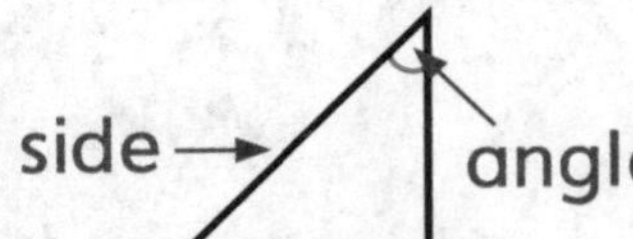

3 sides
3 angles

quadrilateral

4 sides
4 angles

pentagon

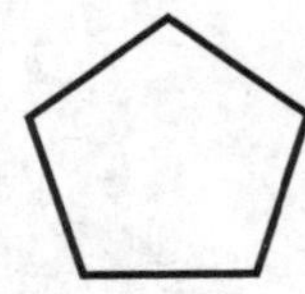

5 sides
5 angles

hexagon

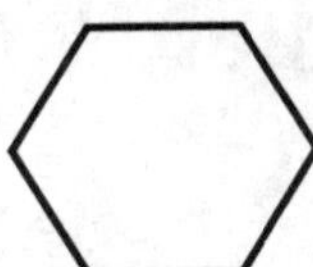

6 sides
6 angles

circle

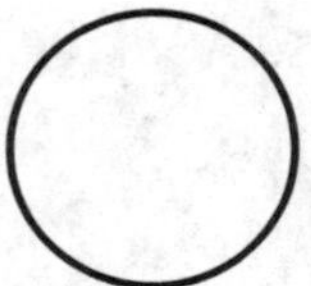

0 sides
0 angles

Practice

Trace each shape. Write how many sides and angles.

1. 

_____ sides

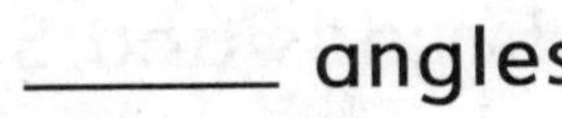

_____ angles

2. 

_____ sides

_____ angles

3. 

_____ sides

_____ angles

4. 

_____ sides

_____ angles

5. Circle the object that has 8 sides and 8 angles.

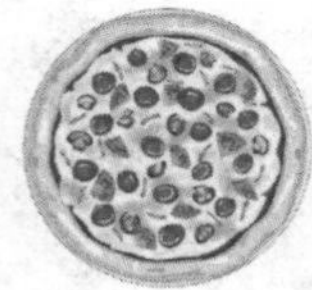

Brain Builders

6. Jason drew a shape that has 6 sides. Petra drew a pentagon. Who drew more sides?

____

7. Carla drew a triangle and a square. Janice drew a shape with 6 sides and 6 angles. Who drew more sides and angles?

____

Vocabulary Check 

Connect the name of each shape to its number of sides or angles.

8. **hexagon**	4 sides and 4 angles
9. **quadrilateral**	5 sides and 5 angles
10. **triangle**	6 sides and 6 angles
11. **pentagon**	3 sides and 3 angles

Math at Home While driving or walking, look at road signs together. Ask your child to name and describe the shapes of the signs he or she sees.

Name ....

Lesson 3

Problem Solving

STRATEGY: Draw a Diagram

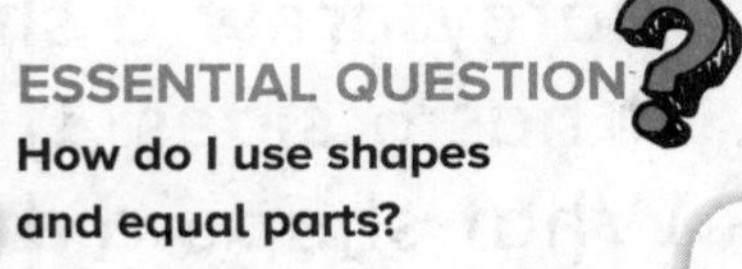

ESSENTIAL QUESTION

How do I use shapes and equal parts?

Lyla drew a shape. The shape has 6 sides. It also has 6 angles. What shape did Lyla draw?

1 Understand

Underline what you know.
Circle what you need to find.

2 Plan

How will I solve the problem?

3 Solve

Draw a diagram.

Lyla drew a ______________.

4 Check

Is my answer reasonable? Explain.

Practice the Strategy

Marcy drew a shape.
It has 5 sides. It has 5 angles.
What shape did she draw?

1 Understand Underline what you know.
Circle what you need to find.

2 Plan How will I solve the problem?

3 Solve I will...

Marcy drew a ________.

4 Check Is my answer reasonable? Explain

Drawing at the park is fun!

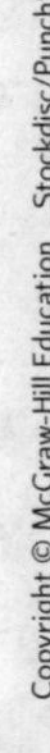

Name ....

Processes &Practices

Apply the Strategy

1. If a shape has 3 sides and 3 angles, what shape is it? Draw the shape.

____

Brain Builders

2. Abby draws a triangle. Samuel draws a shape that has 1 more side than a triangle. What shape did Samuel draw? Draw the shape.

____

3. Jason drew a shape that has more sides than a triangle or rectangle but less angles than a hexagon. What shape did he draw? Draw the shape.

____

Review the Strategies

Choose a strategy
- **Write a number sentence.**
- **Draw a diagram.**
- **Use logical reasoning.**

4. Tammy saw a two-dimensional shape. The shape has 6 sides and 6 angles. Two of the sides are longer than the others. What shape did Tammy see?

____

5. The sign at the end of David's street is a two-dimensional shape. It has 4 sides and 4 angles. The sides are all the same length. What shape is the sign?

____

6. Jason was drawing shapes with sidewalk chalk in the park. He drew 3 triangles and then 2 squares. How many angles did he draw?

_______________ angles

Chalk is cool!

Name

My Homework

Lesson 3

Problem Solving: Draw a Diagram

Jacob was looking for shapes in the stars. He found one with 4 equal length sides and 4 angles. What shape did Jacob see in the stars?

1 Understand Underline what you know. Circle what you need to find.

2 Plan How will I solve the problem?

3 Solve Draw a diagram.

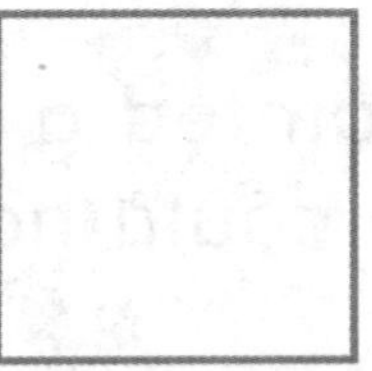

Jacob saw a square.

4 Check Is my answer reasonable? Explain.

Underline what you know. Circle what you need to find. Draw a diagram to solve.

1. Maggie drew a house. She drew a square for the bottom. She drew a triangle on top of the square for the roof. Trace the outside of the house. What shape is Maggie's house? Draw it.

____

Brain Builders

2. Billy saw a sign while walking through the park. The sign had no sides and no angles. Billy says the shape of the sign is a square. Is he correct? Explain.

____

____

____

3. Landon painted a quadrilteral. What are 3 shapes he could have painted?

____

____

____

Math at Home Describe a shape to your child. Have him or her draw the shape you described and identify the shape.

Name

Check My Progress

Vocabulary Check 

Complete each sentence.

angle	**hexagon**	**pentagon**
side	**triangle**	**two-dimensional shape**

1. A ______________ has 5 sides and 5 angles.

2. A ______________ has 3 sides and 3 angles.

3. A ______________ has 6 sides and 6 angles.

Concept Check

Circle the shape or shapes that match the name.

4. triangle

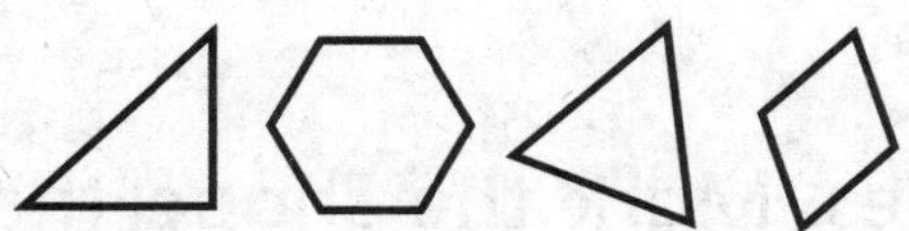

5. pentagon

6. hexagon

7. quadrilateral

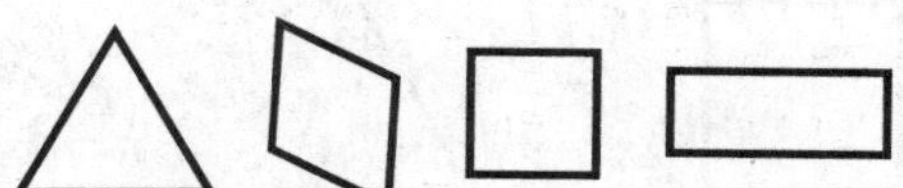

Write the name of the shape. Circle the shapes that match.

8. 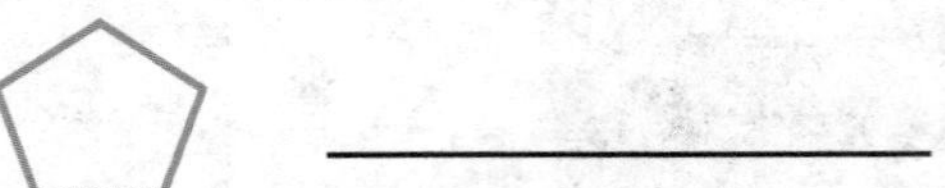 ______________

9.  ______________

Brain Builders

Write how many sides and angles. Circle the shapes that have 1 less side than a pentagon.

10. 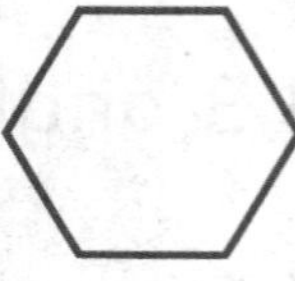

____ sides

____ angles

11. 

____ sides

____ angles

12. 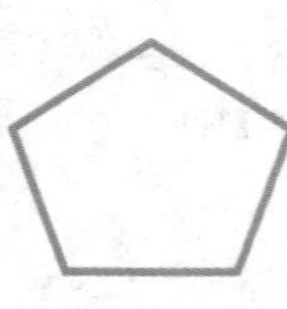

____ sides

____ angles

13. 

____ sides

____ angles

14. **Test Practice** Look at the shapes. Mark the shape that does not belong.

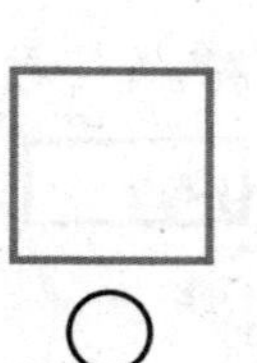 ○

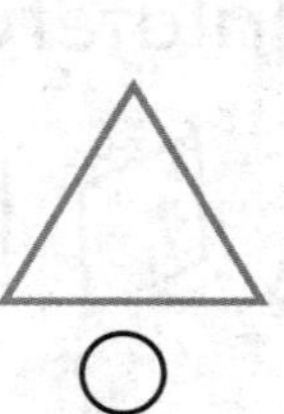 ○

 ○

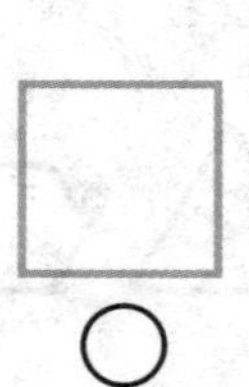 ○

Name

Lesson 4
Three-Dimensional Shapes

ESSENTIAL QUESTION
How do I use shapes and equal parts?

 Math in My World

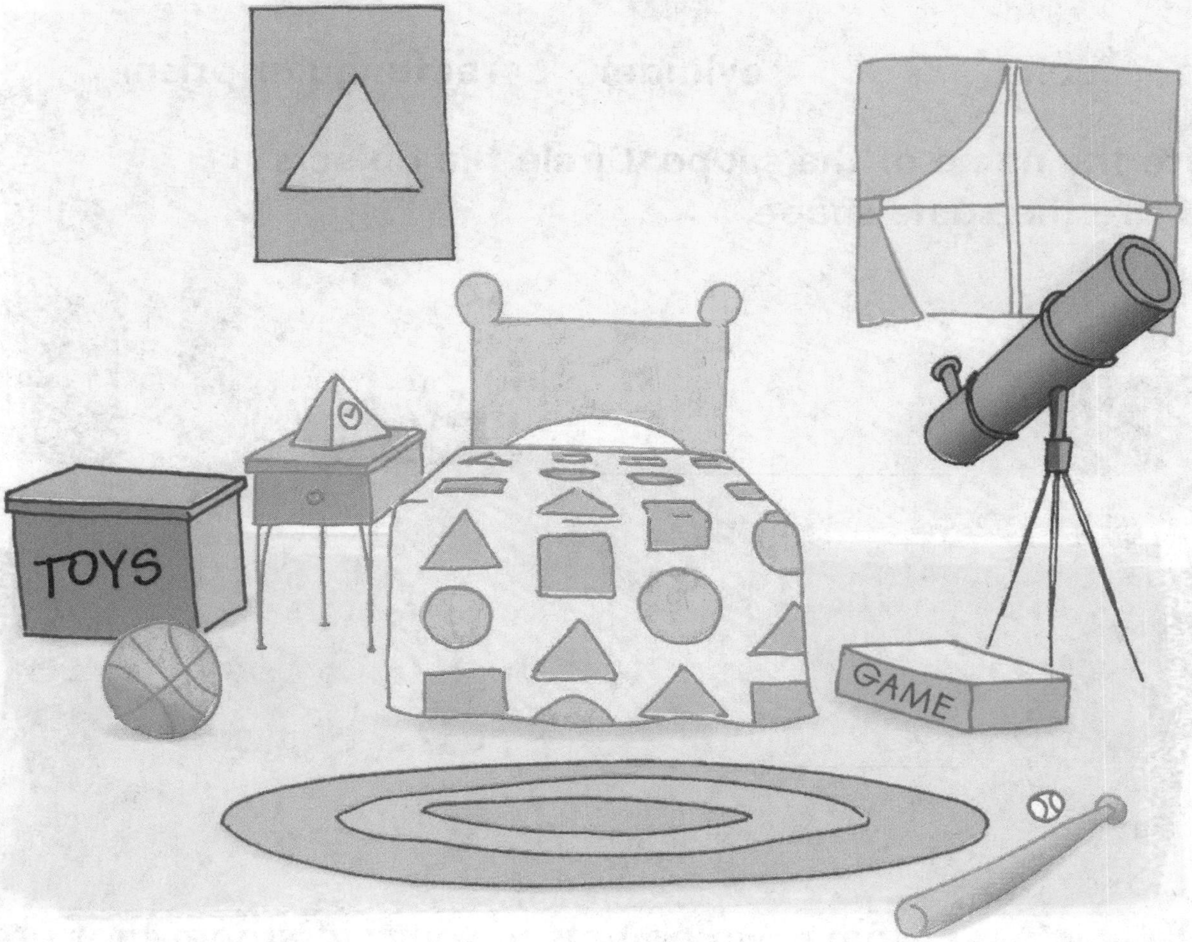

Teacher Directions: Look at the picture. Circle the three-dimensional shapes that you see. Identify and describe each shape.

Processes
&Practices

Guided Practice

A **three-dimensional shape** is a shape with length, width, and height.

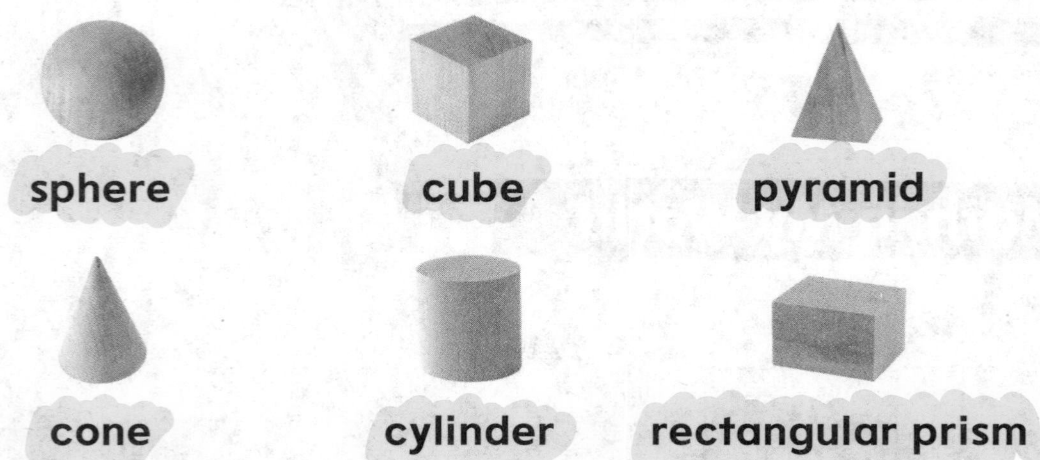

Write the name of the shape. Circle the objects that are the same shape.

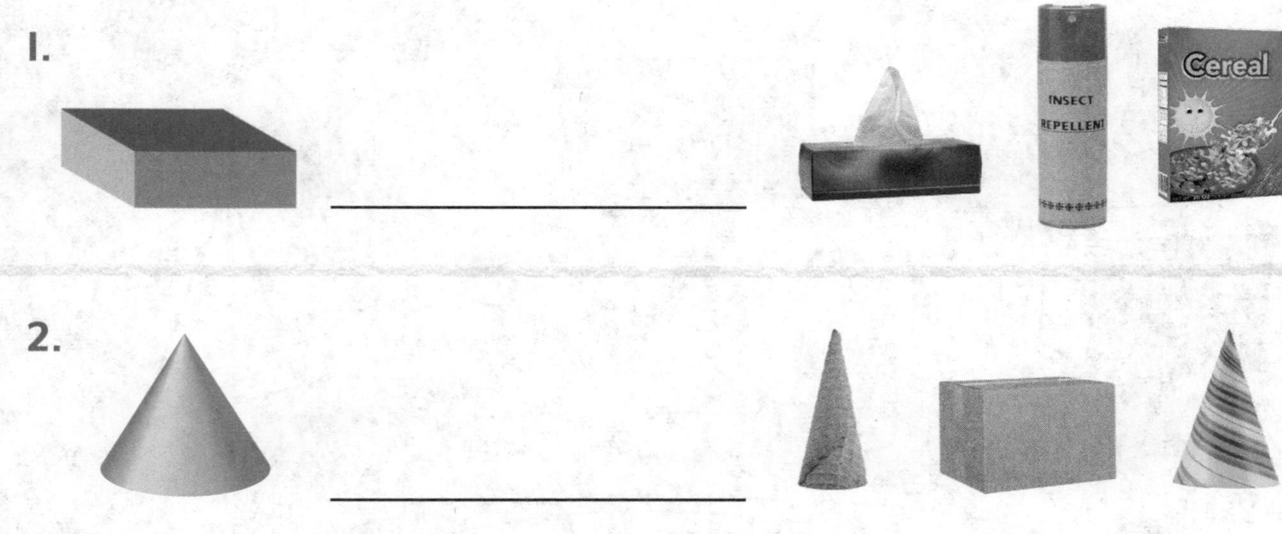

1. ______________

2. ______________

Talk Math Name two objects in your classroom that are the same shape as a rectangular prism.

Name ______________________________

Independent Practice

Write the name of the shape. Circle the objects that are the same shape.

3.

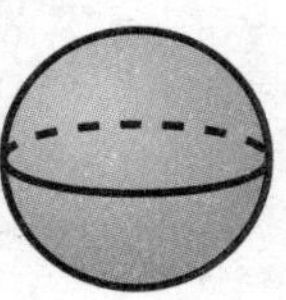

____

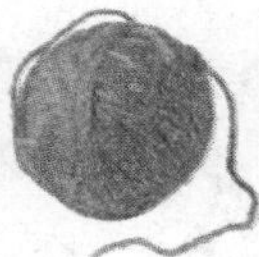

4.

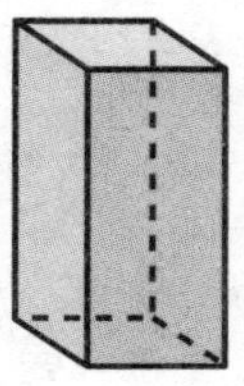

____

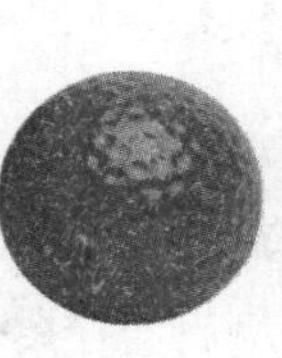

5.

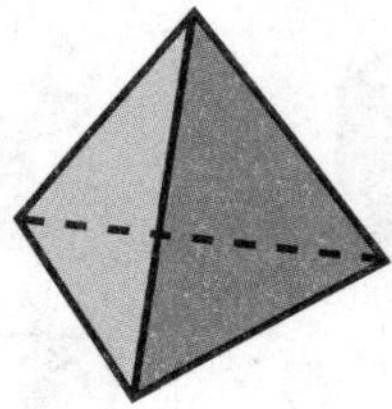

____

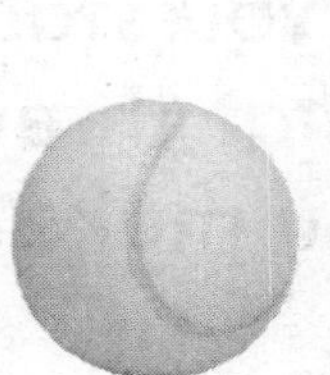

6.

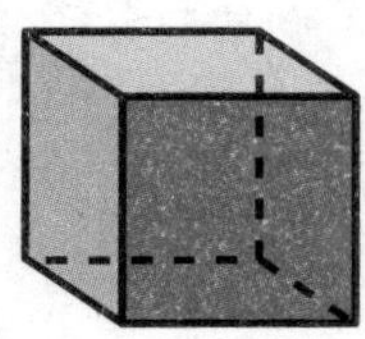

____

7.

____

Problem Solving

8. Jen is wrapping a present that is in a box. The box is square on all sides. What shape is the box Jen is wrapping?

____

9. I am a three-dimensional shape. I have a circle at the bottom. I have a point at the top. What shape am I?

____

Brain Builders

10. If you stack two cubes together what three-dimensional shape will you make?

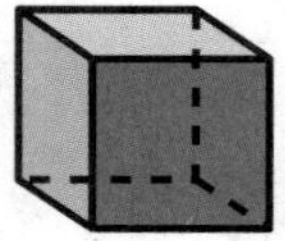

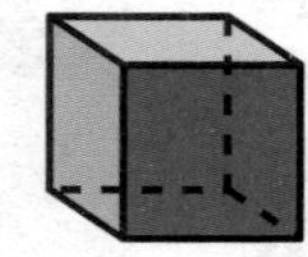

____

Write Math How can you tell if a shape is three-dimensional?

____

____

____

Name ______________________

My Homework

Lesson 4

Three-Dimensional Shapes

Homework Helper

eHelp

Need help? connectED.mcgraw-hill.com

A three-dimensional shape has length, width, and height.

sphere

cube

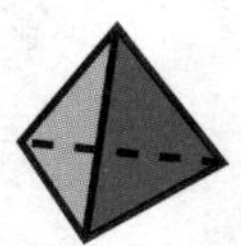
pyramid

cone

cylinder

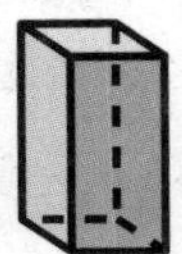
rectangular prism

Practice

Write the name of the shape. Circle the objects that are the same shape.

1. 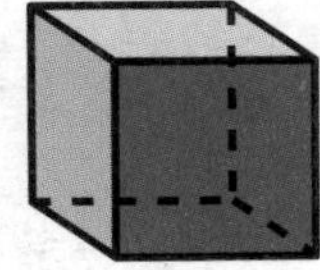 ______________________

2. 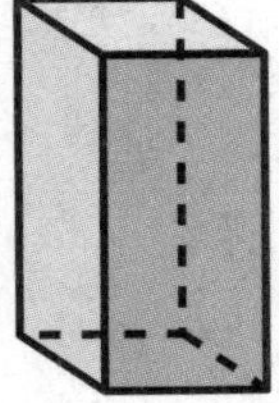 ______________________

3. 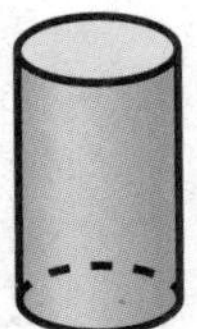 ______________________

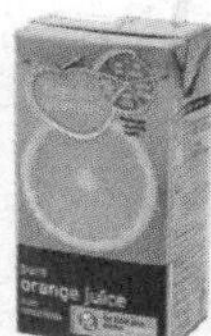
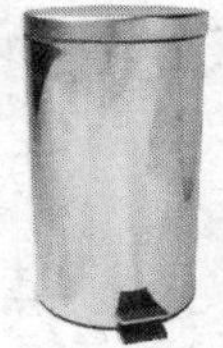

Write the name of each shape. Circle the objects that are the same shape.

4. 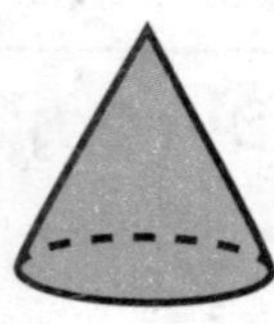 ______________________

5. 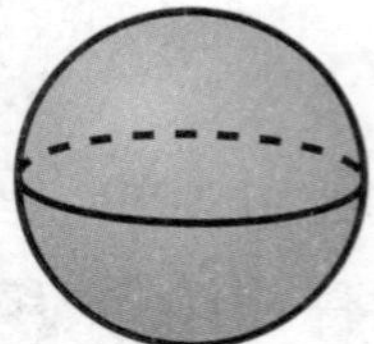 ______________________

Brain Builders

6. I have 6 surfaces. 2 of my surfaces are smaller than the others. I can stand up tall. What shape am I? Explain your answer to a family member or friend.

____

Vocabulary Check

Draw lines to match.

7. **cylinder**

8. **rectangular prism**

9. **cube**

10. **cone**

Math at Home Have your child identify items in your home that match the shapes he or she learned about in this lesson.

Name

Lesson 5

Faces, Edges, and Vertices

ESSENTIAL QUESTION
How do I use shapes and equal parts?

Math in My World

11 10
7
4 3
8 9

6

5
1 2

The shape is a ______________.

Teacher Directions: Start at 1. Connect the dots in number order. Write the name of the shape you drew.

Processes & Practices

Guided Practice

You can describe three-dimensional shapes by the number of faces, edges, and vertices.

edge face vertex

A **face** is a flat surface.

An **edge** is where 2 faces meet.

A **vertex** is where 3 or more faces meet.

Use three-dimensional shapes. Count the faces, edges, and vertices.

Shape	Faces	Edges	Vertices
1. cube	______	______	______
2. rectangular prism	______	______	______
3. pyramid	______	______	______
4. sphere	______	______	______

Talk Math What figure has 6 equal faces? How do you know?

Name

Independent Practice

Circle the shapes that match the description.

5. 0 faces, 0 edges, 0 vertices

6. 6 faces, 12 edges, 8 vertices

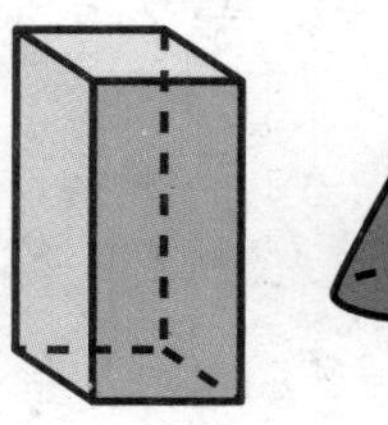  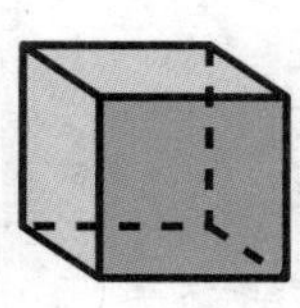

7. 5 faces, 8 edges, 5 vertices

 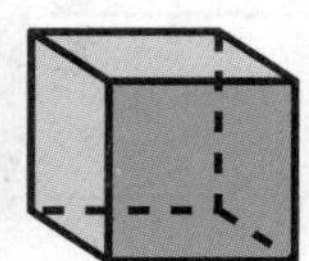 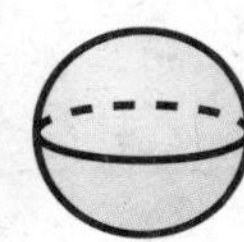

8. 6 equal faces, 12 edges, 8 vertices

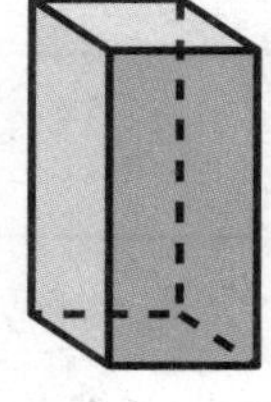 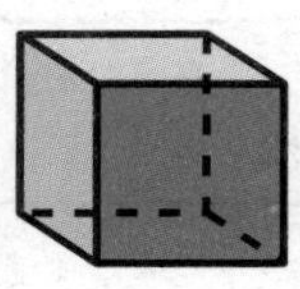 

Circle the objects that match the description.

9. 6 faces, 12 edges, 8 vertices

10. 0 faces, 0 edges, 0 vertices

11. 6 faces, 12 edges, 8 vertices

12. more than 2 faces, more than 0 vertices

  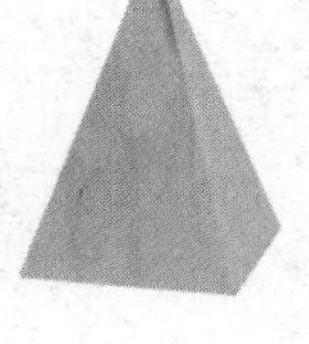

Problem Solving

13. Mindy drew the three shapes below. Circle the shape that has 6 faces and 12 edges.

  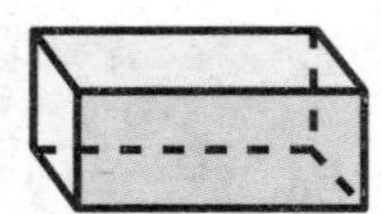

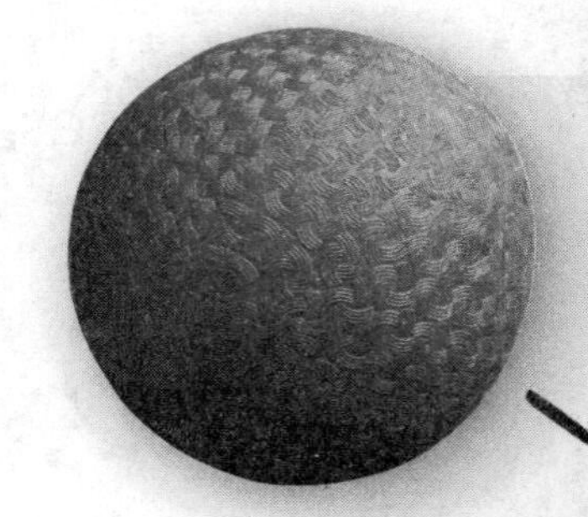

Brain Builders

14. How many faces does a cone have? Does a cone have a vertex? Explain.

____

____

____

____

15. Which shape does not belong? Circle it. Explain why it does not belong.

 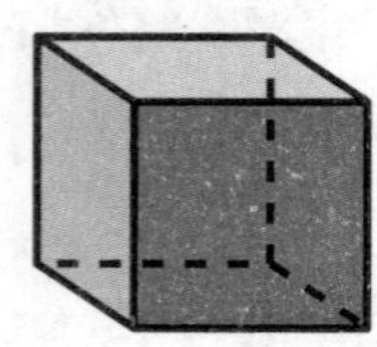 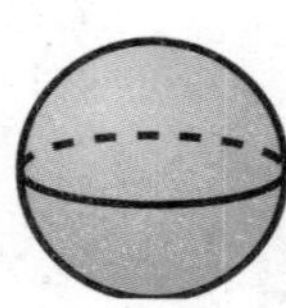 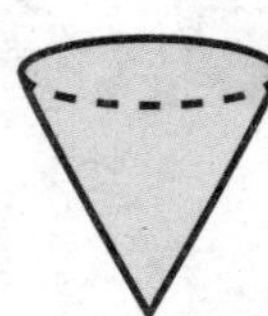

____

____

____

Name ....

My Homework

Lesson 5

Faces, Edges, and Vertices

Homework Helper

Need help? connectED.mcgraw-hill.com

Three-dimensional shapes are described by the number of faces, edges, and vertices.

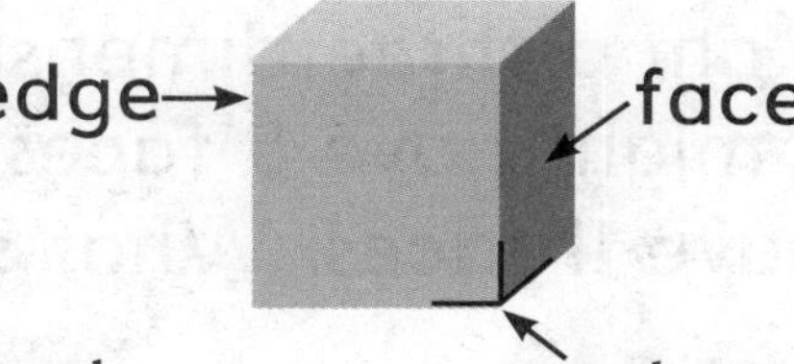

A face is a flat surface.

An edge is where 2 faces meet.

A vertex is where 3 or more faces meet.

Practice

Circle the shapes or objects that match the description.

1. 6 faces, 12 edges, 8 vertices

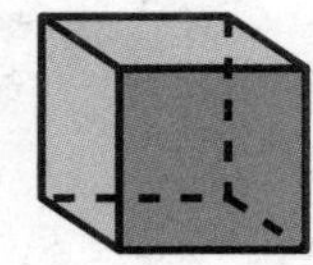 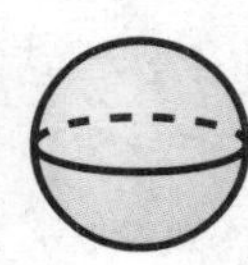 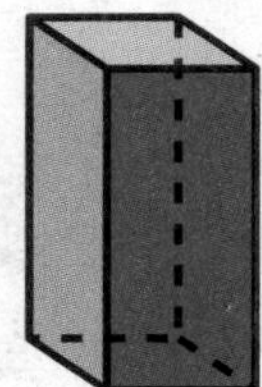

2. 0 faces, 0 edges, 0 vertices

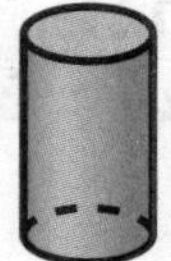  

3. 5 faces, 8 edges, 5 vertices

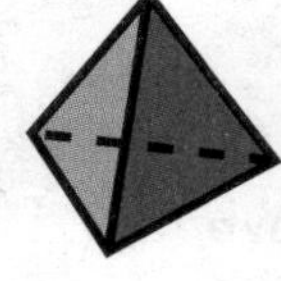 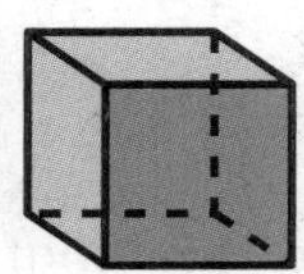 

4. 6 faces, 12 edges, 8 vertices

Circle the objects that match the descriptions.

5. 6 faces, 12 edges, 8 vertices

6. 0 faces, 0 edges, 0 vertices

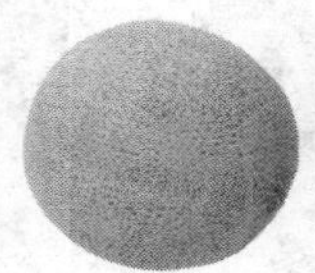

Brain Builders

7. I am a three-dimensional shape. I might have 5 faces. Or I might have 4 faces. What shape am I? Explain.

____

____

____

Vocabulary Check 

Complete each sentence.

face **edge** **vertex**

8. A __________ is a flat surface.

9. A __________ is where 3 or more faces meet.

10. An __________ is where 2 faces meet.

Math at Home Have your child identify real-life objects in your home that have the same shape as one of the shapes learned in this lesson.

Name

Lesson 6
Relate Shapes and Solids

ESSENTIAL QUESTION
How do I use shapes and equal parts?

Math in My World

Teacher Directions: Trace one face of a cube. Identify the shape. Trace the other faces of the cube. Describe the faces of a cube.

Guided Practice

The faces of three-dimensional shapes are two-dimensional shapes.

Helpful Hint
A cube has 6 equal faces. The faces are squares.

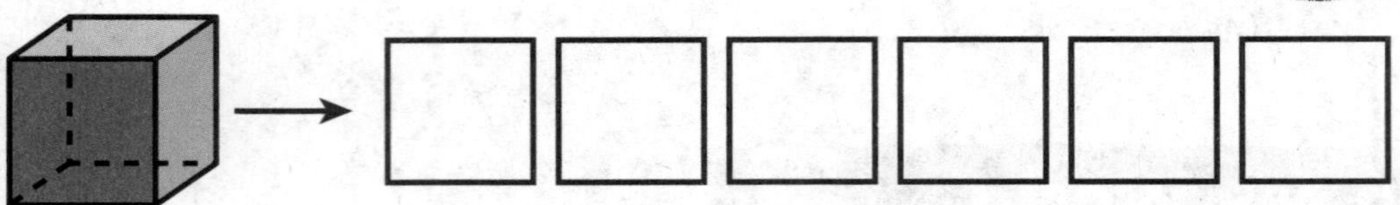

Circle the faces that make the shape.

1. 

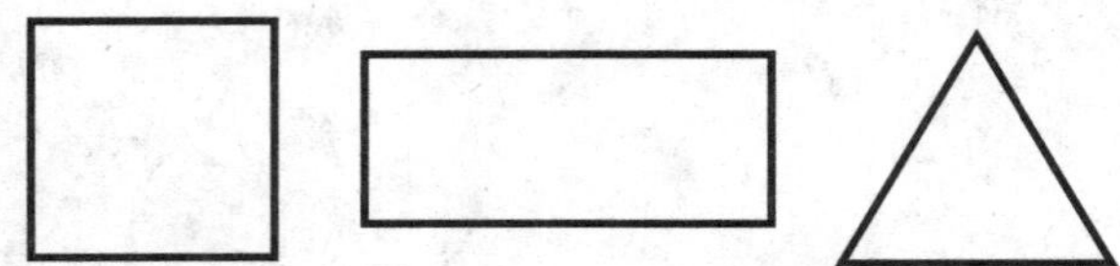

2. 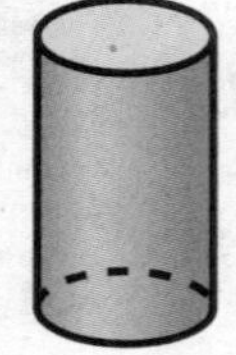

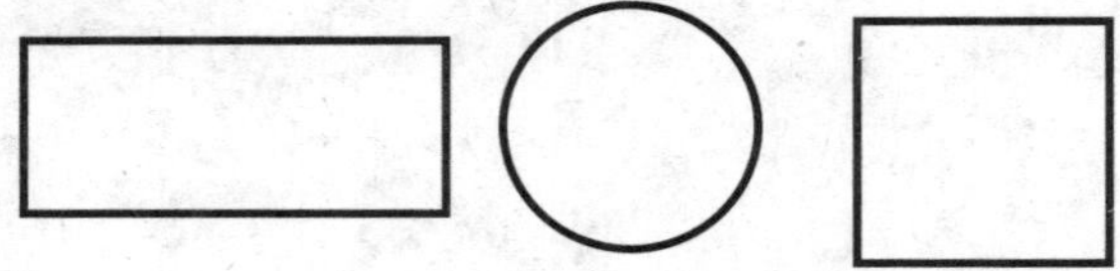

3. 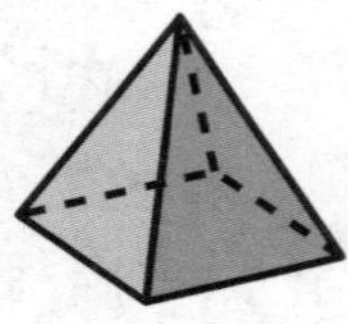

Talk Math Explain how two-dimensional shapes and three-dimensional shapes are related.

Name

Independent Practice

Circle the faces that make the shape.

4. 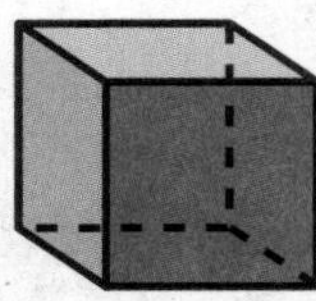

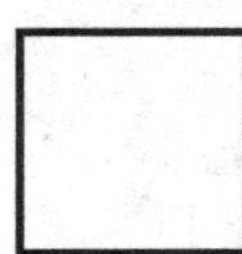 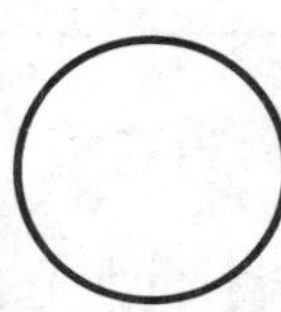 

5. 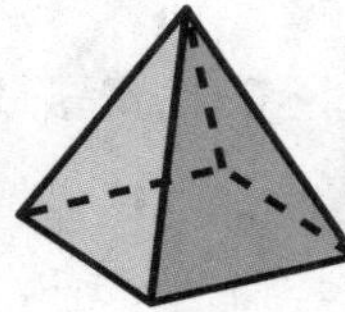

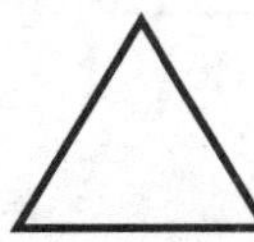 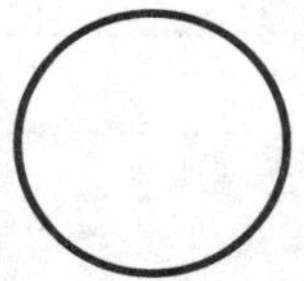 

Circle the shape made by the faces.

6. 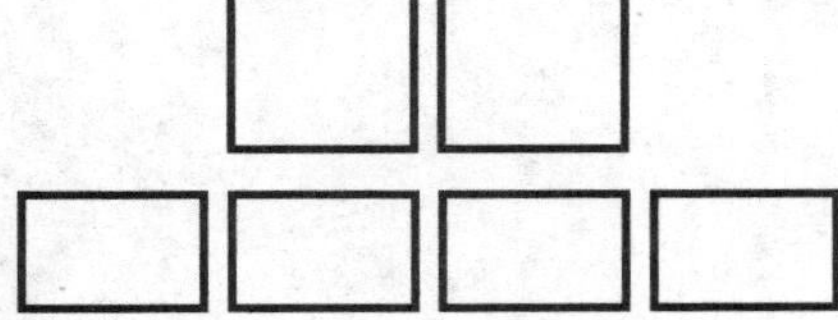

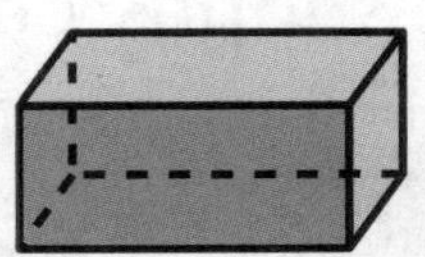 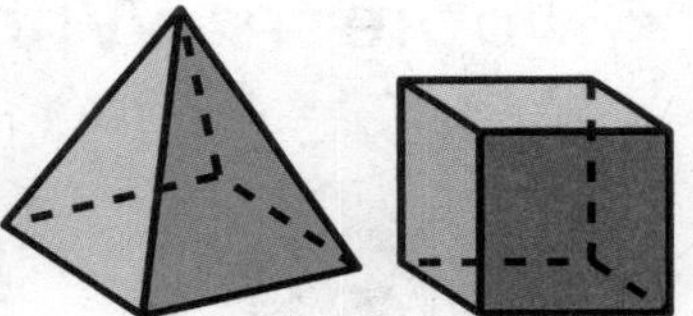

7. 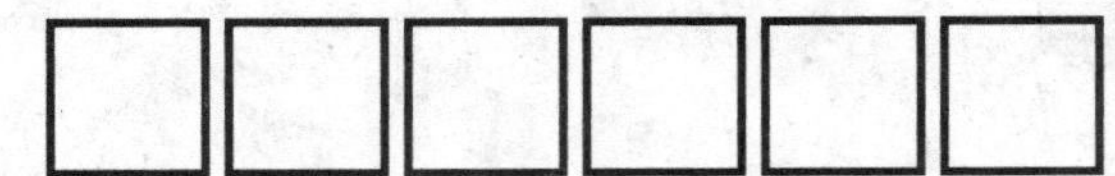

8. Which of these shapes does not have a square as one of its faces?

 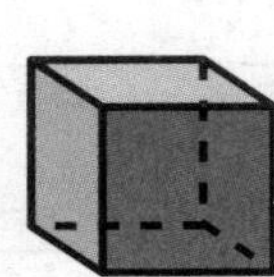 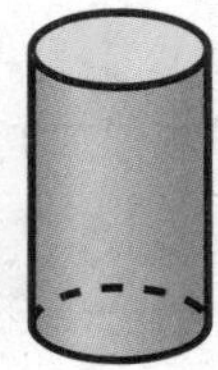 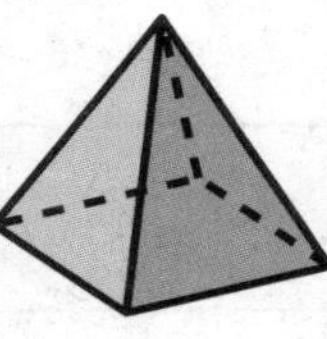

Problem Solving

9. I have 6 equal faces. I have 8 vertices. What shape am I?

____

10. I have no faces and no vertices. What shape am I?

____

Brain Builders

11. Alison wants to trace shapes using these objects. What shapes can she trace?

____

Write Math Describe the faces that make a pyramid.

____

____

____

Name ....

My Homework

Lesson 6

Relate Shapes and Solids

Homework Helper

Need help? connectED.mcgraw-hill.com

The faces of three-dimensional shapes are two-dimensional shapes.

Practice

Circle the faces that make the shape.

1.  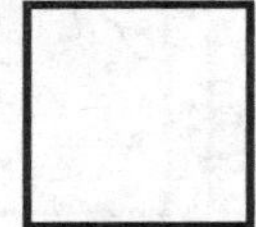 

2. 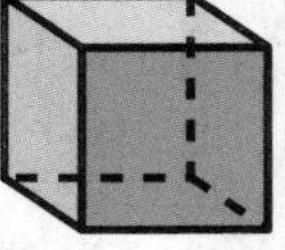  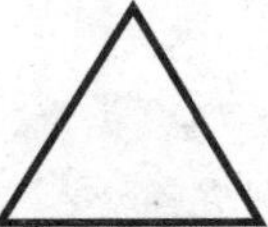 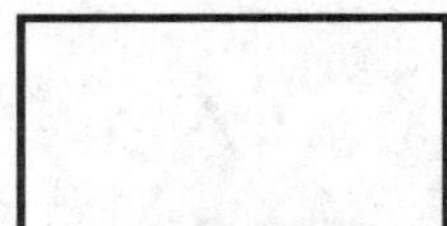

3. 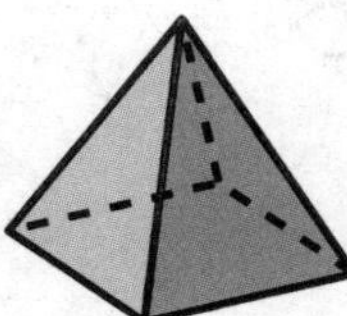 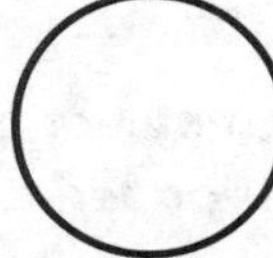  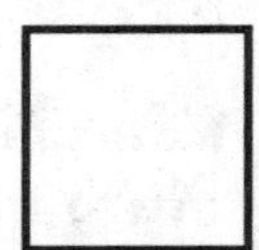

Circle the shape made by the faces.

4.

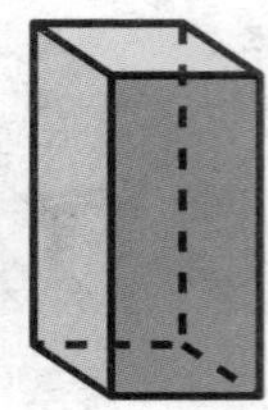

5.

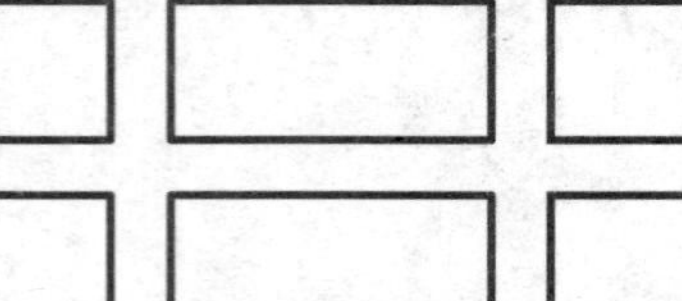

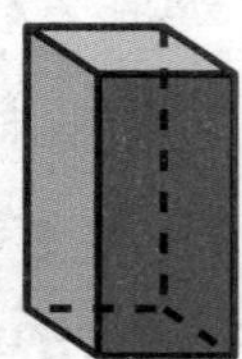

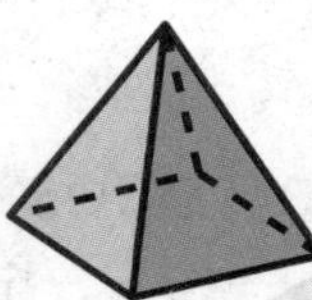

Brain Builders

6. If you put these shapes together, what three-dimensional shape could you make? Write the name of the shape.

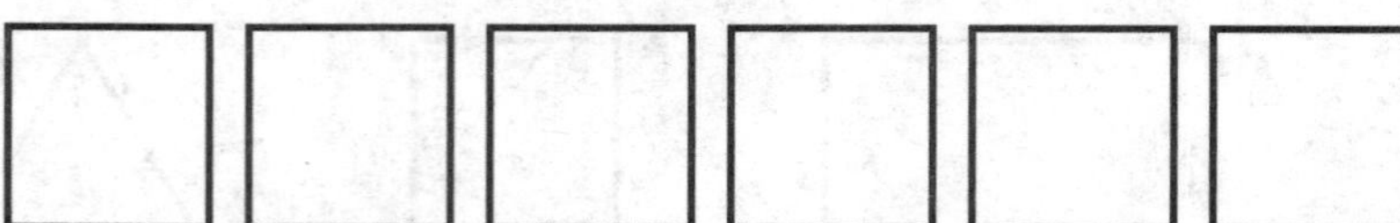

____

7. **Test Practice** Identify the shape that does not belong.

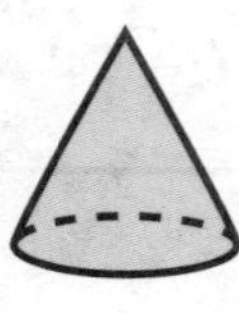

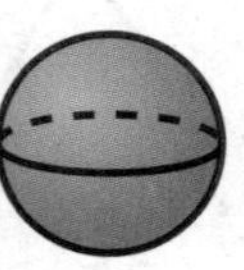

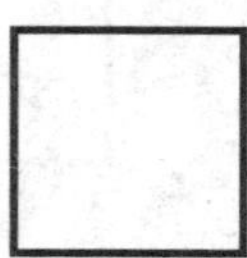

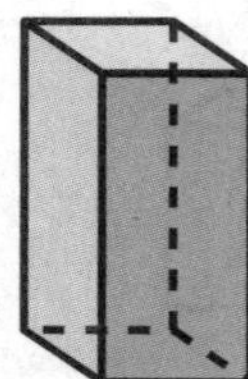

Math at Home Ask your child to find an object at home that he or she could use to trace a rectangle on a piece of paper. Challenge your child to see if he or she can find something to trace for a circle.

Name

Lesson 7
Halves, Thirds, and Fourths

ESSENTIAL QUESTION
How do I use shapes and equal parts?

 Math in My World 

Picnic!

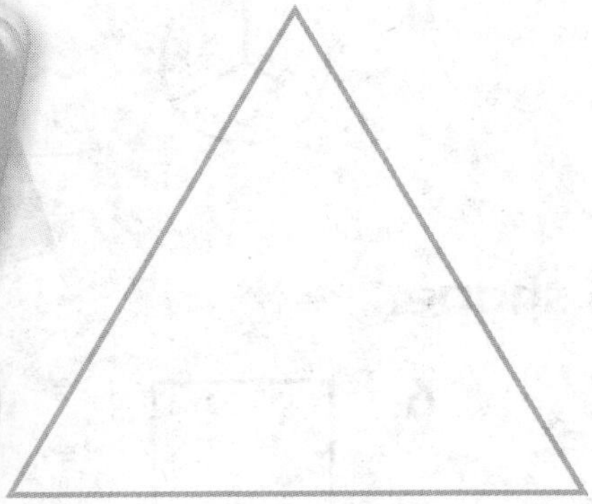

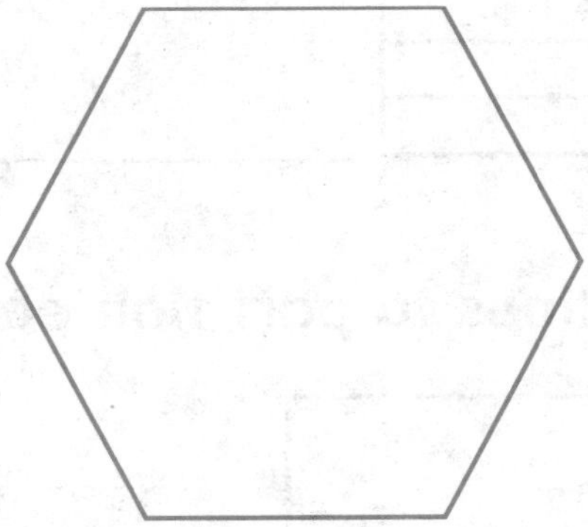

Teacher Directions: Use square, triangle, and trapezoid pattern blocks to cover each shape. Trace the blocks to show the shapes you used. Write how many blocks you used to cover each shape.

Guided Practice

You can **partition**, or separate, shapes into equal parts.

Two equal parts or two **halves.**
Each part is **half of** the whole.

1
2

Three equal parts or three **thirds.**
Each part is a **third of** the whole.

1
2
3

Four equal parts or four **fourths.**
Each part is a **fourth of** the whole.

1 2
3 4

1 2
3 4

Describe the equal parts. Write *two halves, three thirds,* or *four fourths.*

1. 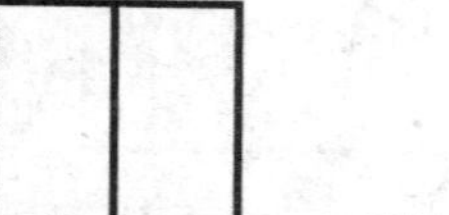

2. 

3. 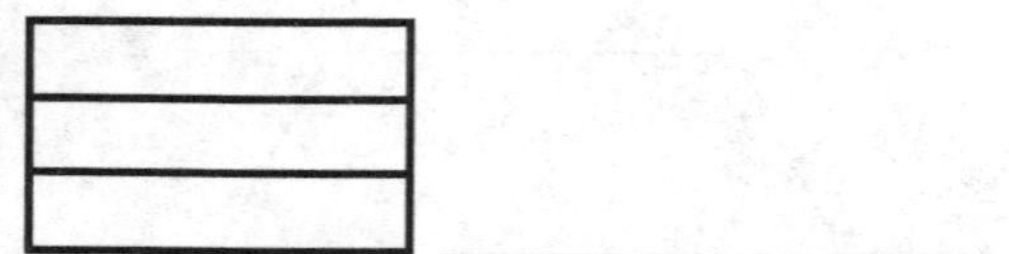

4. 

Draw lines to partition each shape.

5. 

2 equal parts

6. 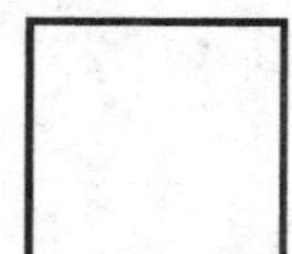

4 equal parts

Talk Math Explain how you can divide a pie so that four people each get an equal part.

Name

Independent Practice

Describe the equal parts. Write *two halves, three thirds,* or *four fourths*.

7. ________________

8. ________________

Draw lines to partition each shape.

9. 4 equal parts

10. 2 equal parts

11. 3 equal parts

12. 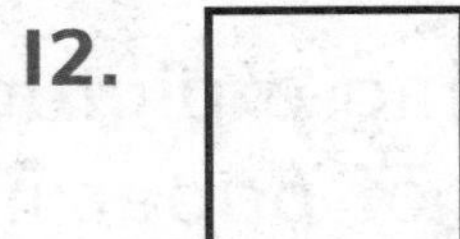

2 equal parts

Partition the shape in a different way. Show the same number of equal parts.

13. 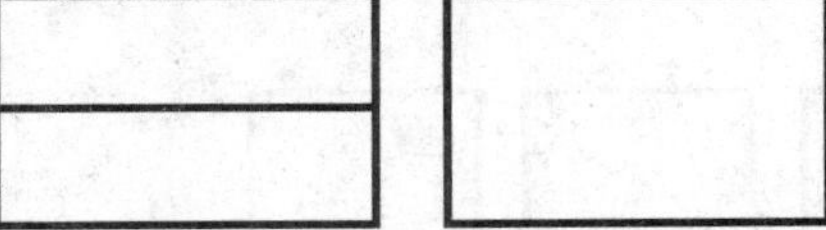

14. 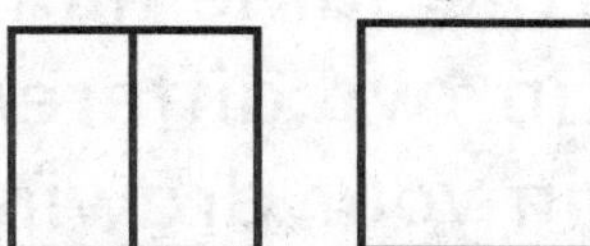

15. 

16. 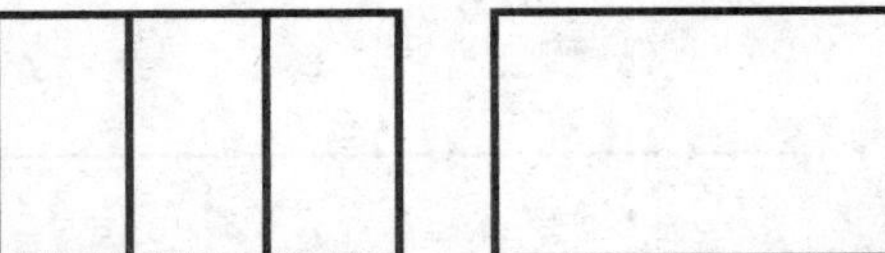

Problem Solving

17. Eva's mom bought a pizza. Eva ate one equal part. Her friend ate one equal part. There was one equal part left for Eva's mom. How much of the pizza was left for Eva's mom?

_______________ of the pizza

18. Gracie had a round slice of watermelon. She and her sister shared the slice equally. How much did each girl eat?

_______________ of the watermelon slice

Brain Builders

19. Sadie is making a picture for her cousin. She folds a piece of paper in half. Then she folds it in half again. She opens the paper. How many equal parts are there?

____

20. Show the same number of equal parts in two different ways. Explain your drawing.

____

____

____

Name ____________________

My Homework

Lesson 7

Halves, Thirds, and Fourths

Homework Helper

Need help? connectED.mcgraw-hill.com

You can partition, or separate, shapes into equal parts.

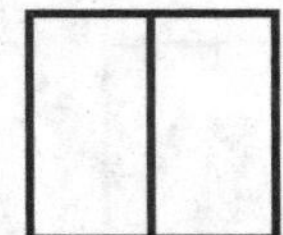

halves

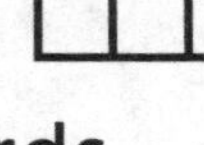

thirds

fourths

Practice

Describe the equal parts. Write *two halves, three thirds, or four fourths.*

1. 

____

2. 

____

3. 

____

4. 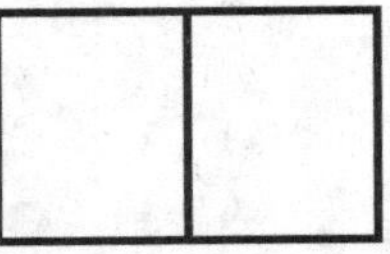

____

Draw lines to partition each shape.

5. 3 equal parts

6. 2 equal parts

7. 4 equal parts

Partition the shape in a different way. Show the same number of equal shares.

8.

9.

Brain Builders

10. Nora cut her sandwich into fourths. How many friends can share her sandwich and have equal parts?

_______________ friends

Vocabulary Check

Vocab

Color each shape as described.

11. **one half** green

12. **one fourth** blue

13. **one third** red

Math at Home Cut your child's food into either halves, thirds, or fourths. Ask him or her to identify how many equal parts you have created.

Name

Lesson 8
Area

ESSENTIAL QUESTION
How do I use shapes and equal parts?

Math in My World

Teacher Directions: Use color tiles to cover each shape. Draw lines to show how the squares fit together to make the shape. Write how many tiles you used.

Processes &Practices

Guided Practice

Rectangles can be partitioned into equal-sized squares. You can count the squares to describe its size.

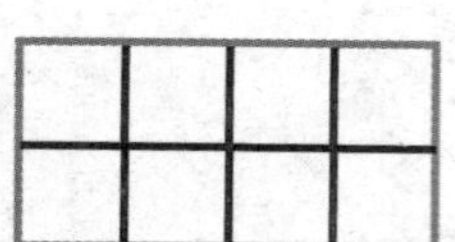 8 squares  12 squares

Count the squares. Write how many squares make each rectangle.

1. 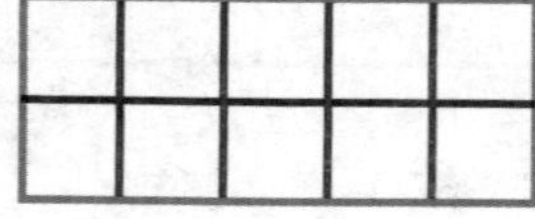

_____ squares

2. 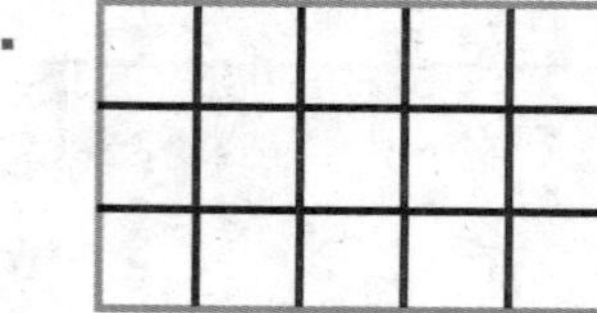

_____ squares

3. 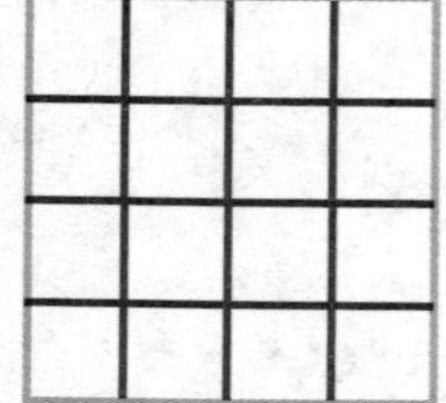

_____ squares

4. 

_____ squares

5. 

_____ squares

6. 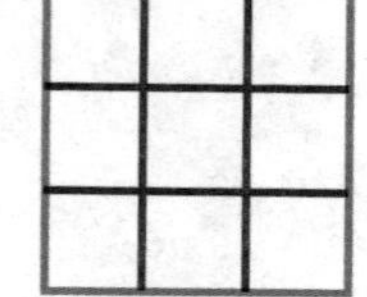

_____ squares

Talk Math Explain how you would partition a rectangle into 6 equal-sized squares.

Name

Independent Practice

Count the squares.
Write how many squares make each rectangle.

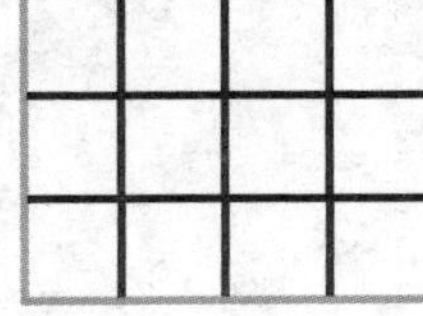

_____ squares

8.

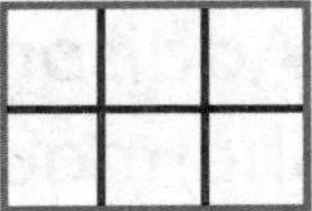

_____ squares

Use color tiles to cover each rectangle.
Write how many tiles were used.

9.

_____ tiles

10.

_____ tiles

Brain Builders

Draw a picture to solve.

11. Addison and Corin each have a group of dominoes. Addison places 4 dominoes in each row. She made 5 rows. Corin places 5 dominoes in each row. She made 3 rows. Who has more dominoes?

____

12. Larry was placing crackers on a tray. The tray held 4 rows of crackers. There were 16 crackers on the plate in all. How many crackers were in each column?

_______________ crackers

Write Math Explain two ways to partition a rectangle into 4 equal parts.

____

____

____

Name

My Homework

Homework Helper

eHelp

Need help? connectED.mcgraw-hill.com

A rectangle can be partitioned into squares to describe its size.

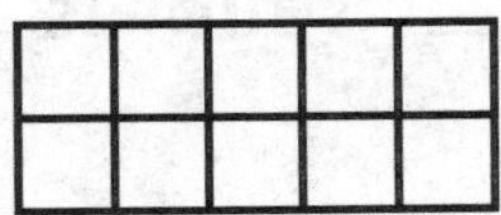

This rectangle is partitioned into 10 squares.

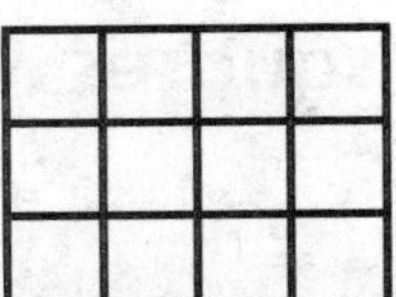

This rectangle is partitioned into 12 squares.

Practice

Count the squares. Write how many squares make each rectangle.

1. 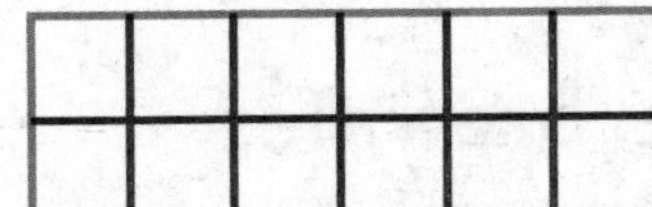

_______ squares

2. 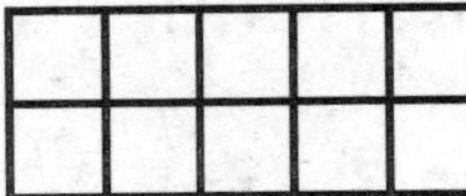

_______ squares

3. 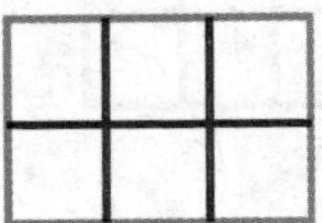

_______ squares

4. 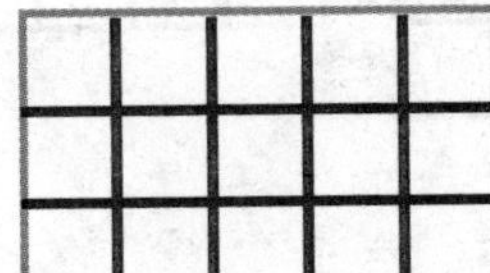

_______ squares

Count the squares. Write how many squares make each rectangle.

5. 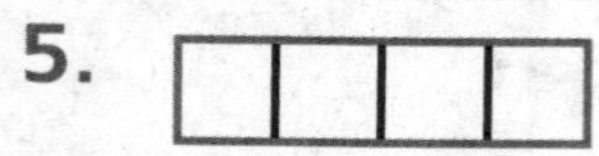

_______ squares

6. 

_______ squares

7. 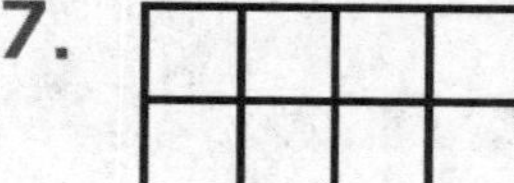

_______ squares

Brain Builders

8. Jan is cutting a rectangular pan of brownies. She cut them in half and then cut each of those halves in half again. She did the same thing going the other direction. How many brownies does she have? Explain your answer to a family member or friend.

_______ brownies

9. **Test Practice** Choose the rectangle below that is partitioned into the greatest number of squares.

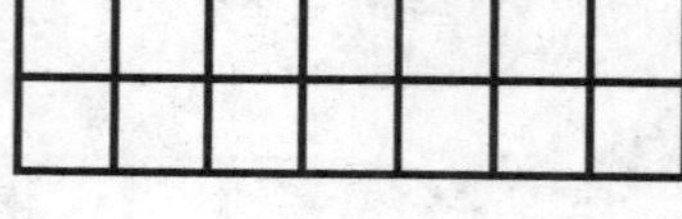
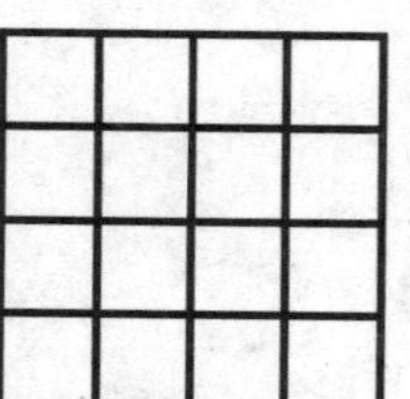
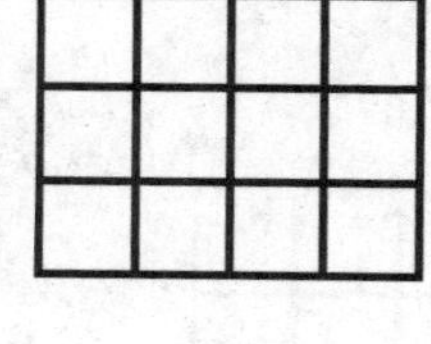

Math at Home Find an opportunity to have your child help you determine how to cut something you are going to serve such as a casserole, brownies, cake, or rice treats. Together determine how many equal pieces you want and how to cut it.

Name ....

My Review

Vocabulary Check

Draw each shape.

1. hexagon

2. quadrilateral

Match each word to the correct shape.

3. **cone**

4. **pyramid**

5. **sphere**

6. **cube**

7. **cylinder**

8. **rectangular prism**

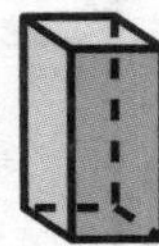

Concept Check

Circle the shapes that match the name.

9. rectangle

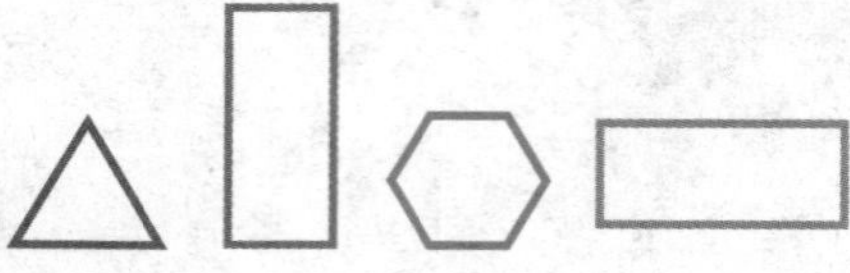

10. pentagon

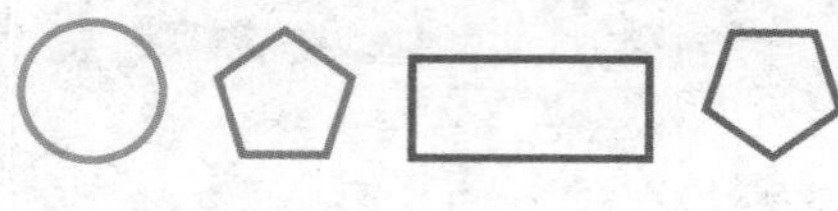

Circle the shape that matches the description.

11. 3 sides and 3 angles

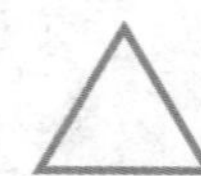

12. 6 sides and 6 angles

Write the name of the shape. Color the shapes that match.

13. 

____

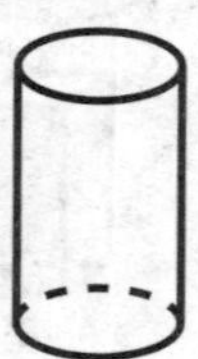
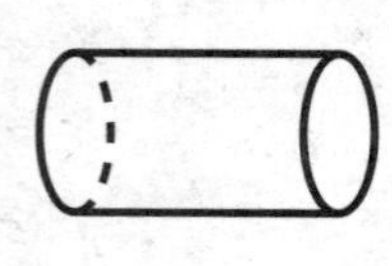

Circle the object that matches the description.

14. 6 faces, 12 edges, 8 vertices

Name ....

Problem Solving

15. Caden's family is having pizza for dinner. There are 4 people in Caden's family. Draw lines to show how Caden's family should cut the pizza so that everyone gets an equal part.

Brain Builders

16. Jamie drew a rectangle. He wants to partition it into equal-sized squares. He starts to partition the rectangle. Finish partitioning his rectangle. Write the total number of squares. Explain your answer to a friend.

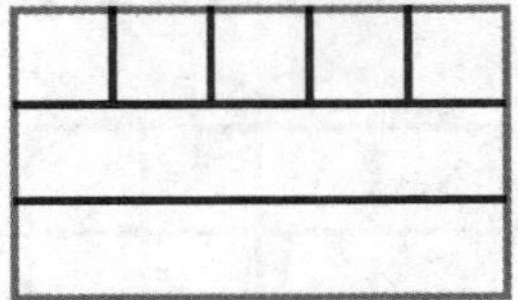

______ squares

Lunchtime!

17. **Test Practice** Blake has a peanut butter and jelly sandwich for lunch. He wants to share it with his 2 friends. Circle the word that tells how Blake should partition his sandwich.

halves	thirds	fourths	fifths
○	○	○	○

Reflect

Chapter 12

Answer the essential question.

Show the ways you can use shapes and equal parts.

ESSENTIAL QUESTION

How do I use shapes and equal parts?

Identify each shape.

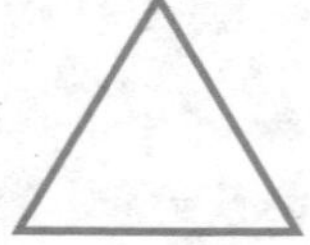

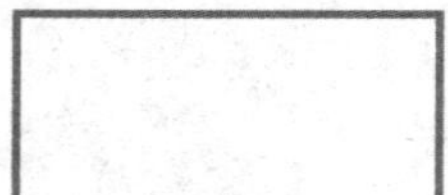

__________ __________

Identify each shape.

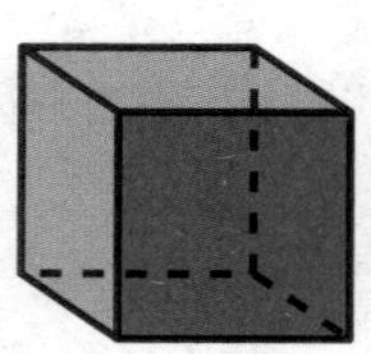

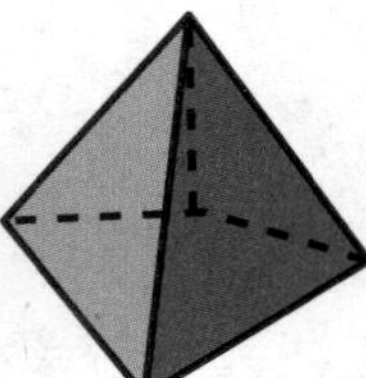

__________ __________

Partition the shape to match the description.

thirds

Write how many squares make the rectangle.

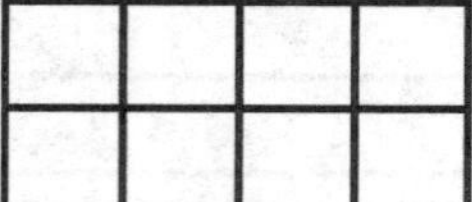

______ squares

Get the swing of it!

Name Date

Score

Performance Task

Shapes in Art

You want to create some designs made of the different shapes shown below. You can use circles, squares, triangles, rectangles, trapezoids, hexagons, and pentagons.

Trapezoid

Rectangle

Pentagon

Hexagon

Circle

Triangle

Square

Show all your work to receive full credit.

Part A

You decide to make the shapes shown using other shapes. For your first design, you use only squares to make your shapes. Which of the shapes are possible to create using only squares? Make a drawing to explain.

Part B

You want more shapes in your design, so you decide to also use triangles to make the shapes. What shapes can you make using only triangles? Make a drawing to explain.

Part C

Are there any shapes from the list that are not possible to make using squares or triangles? Explain.

Glossary/Glosario

Aa English | Spanish/Español

A.M. The hours from midnight until noon.

a.m. Las horas que van desde la medianoche hasta el mediodía.

add (addition) Join together sets to find the total or sum. The opposite of *subtract*.

$2 + 5 = 7$

sumar (adición) Unir conjuntos para hallar el total o la suma. Lo opuesto de restar.

$2 + 5 = 7$

addend Any numbers or quantities being added together.

In $2 + 3 = 5$, 2 is an addend and 3 is an addend.

$2 + 3 = 5$
↑ ↑

sumando Numeros o cantidades que se suman.

En $2 + 3 = 5$, 2 es un sumando y 3 es un sumando.

$2 + 3 = 5$
↑ ↑

Aa

after Follow in place or time.

5678

6 is just *after* 5

después Que sigue en lugar o en tiempo.

5678

6 está justo *después* del 5

analog clock A clock that has an hour hand and a minute hand.

reloj analógico Reloj que tiene una manecilla horaria y un minutero.

angle Two sides on a two-dimensional shape meet to form an angle.

ángulo Dos lados de una figura bidimensional se encuentran para formar un ángulo.

array Objects displayed in rows and columns.

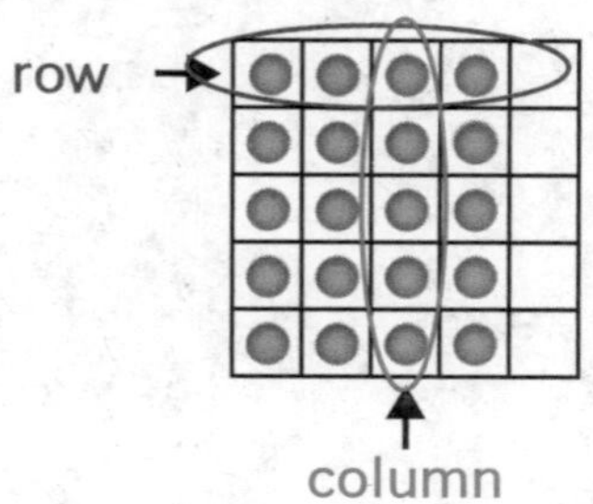

arreglo Objetos organizados en filas y columnas.

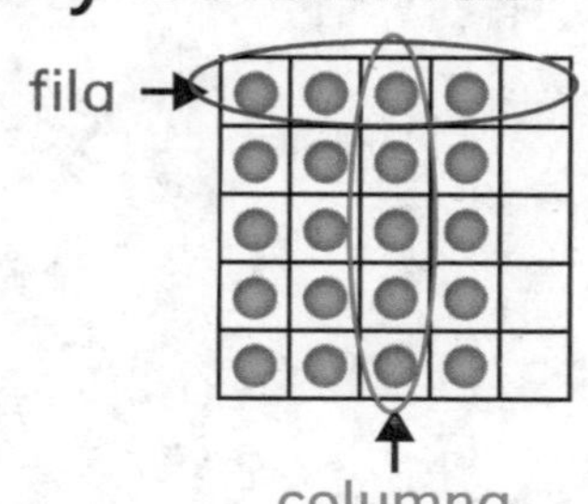

Bb

bar graph A graph that uses bars to show data.

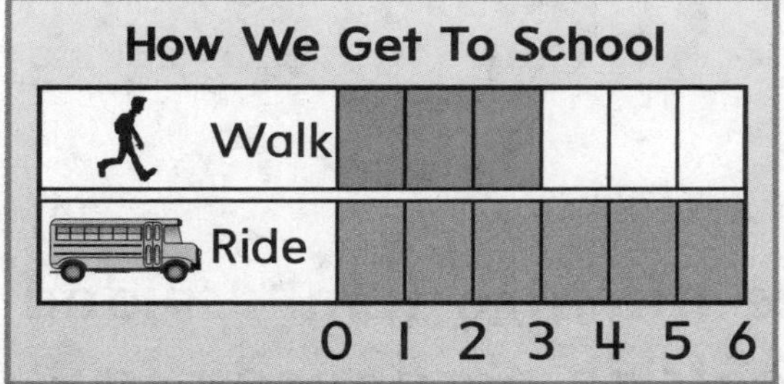

gráfica de barras Gráfica que usa barras para ilustrar datos.

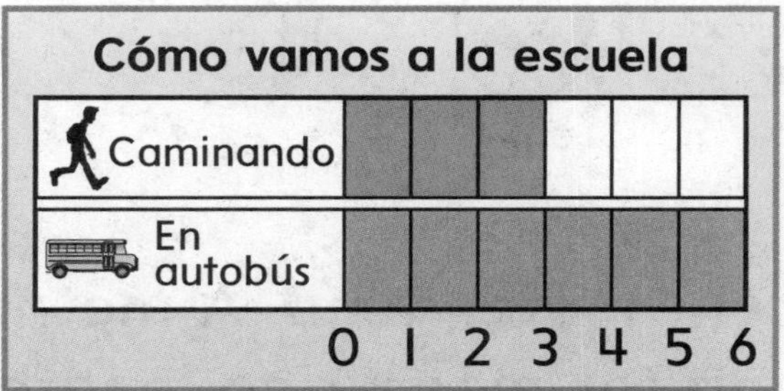

before

5 6 7 8

6 is just *before* 7

antes

5 6 7 8

6 está justo *antes* del 7

between

47 48 49 50

49 is *between* 48 and 50

entre

47 48 49 50

49 está *entre* 48 y 50

Cc

cent

1¢ 1 cent

centavo

1¢ 1 centavo

Cc

cent sign (¢) The sign used to show cents.

1¢

5¢

signo de centavo (¢) El signo que se usa para mostrar centavos.

1¢

5¢

centimeter (cm) A metric unit for measuring length.

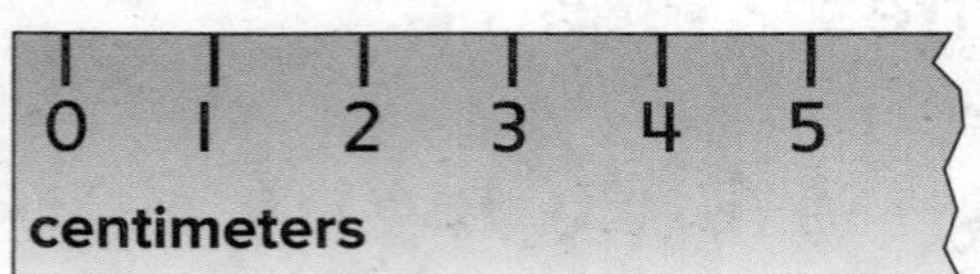

centímetro (cm) Unidad métrica para medir la longitud.

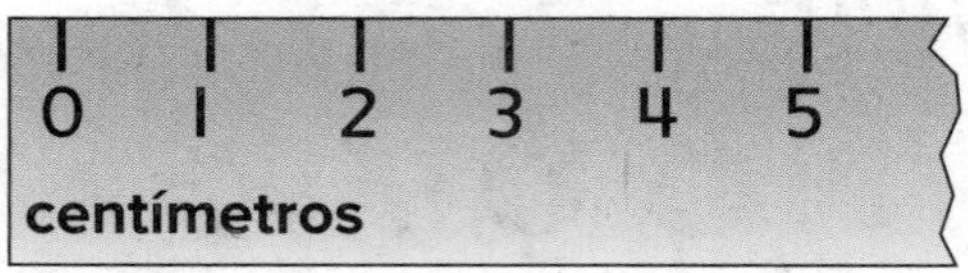

circle A closed, round two-dimensional shape.

círculo Bidimensional redonda y cerrada.

compare Look at objects, shapes, or numbers and see how they are alike or different.

comparar Observar objetos, formas o números para saber en qué se parecen y en qué se diferencian.

cone A three-dimensional shape that narrows to a point from a circular face.

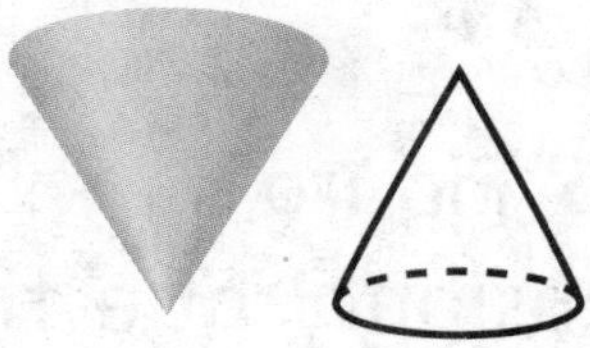

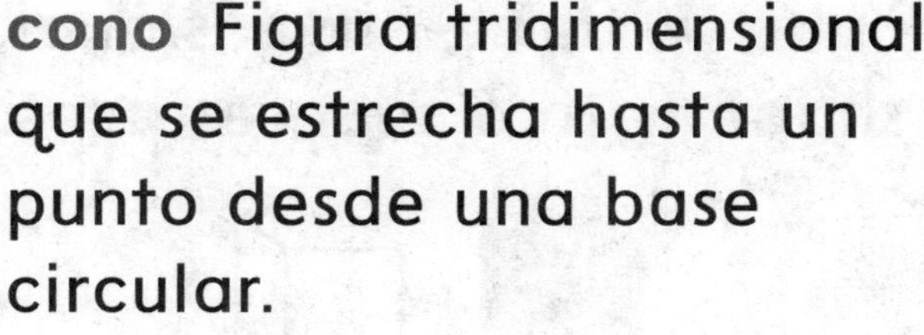

cono Figura tridimensional que se estrecha hasta un punto desde una base circular.

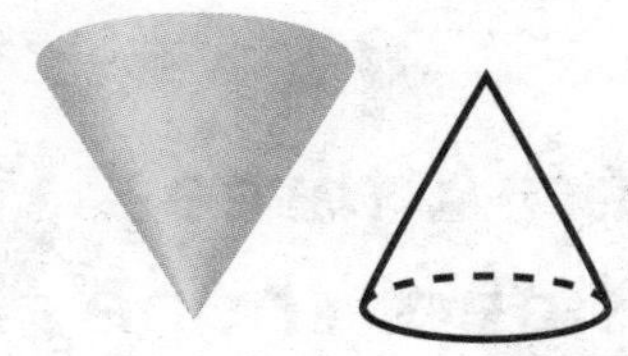

count back On a number line, start at the greater number (5) and count back (3).

$5 - 3 = 2$

2 3 4 5 6

contar hacia atrás En una recta numérica, comienza en un número mayor (5) y cuenta (3) hacia atrás.

$5 - 3 = 2$

2 3 4 5 6

count on On a number line, start at the first addend (4) and count on (2).

$4 + 2 = 6$

3 4 5 6 7

seguir contando En una recta numérica, comienza en el primer sumando (4) y cuenta (2) hacia delante.

$4 + 2 = 6$

3 4 5 6 7

cube A three-dimensional shape with 6 square faces.

cubo Figura tridimensional con 6 caras cuadradas.

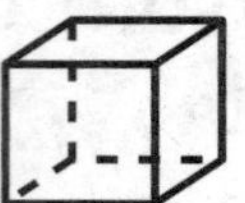

cylinder A three-dimensional shape that is shaped like a can.

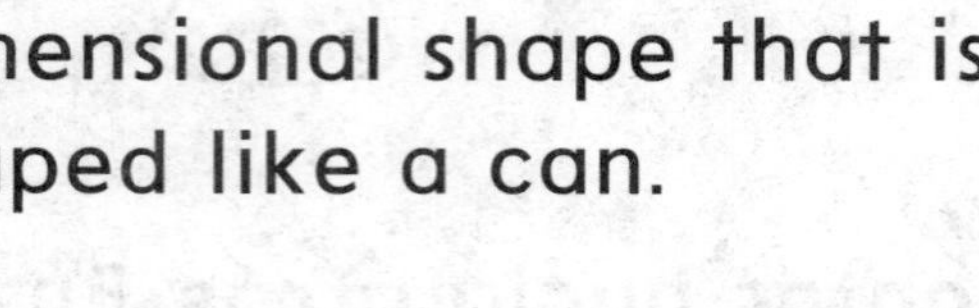

cilindro Figura tridimensional que tiena la forma de una lata.

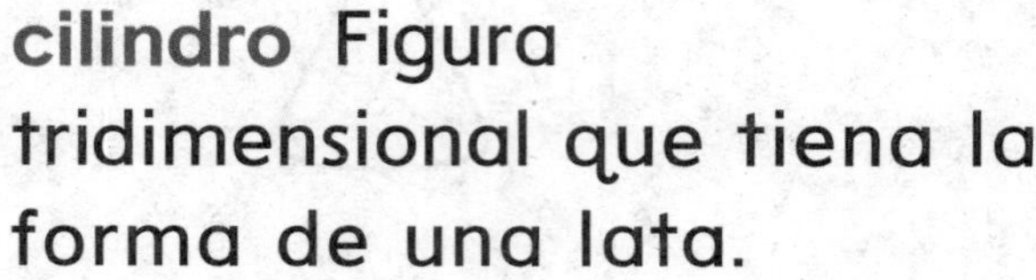

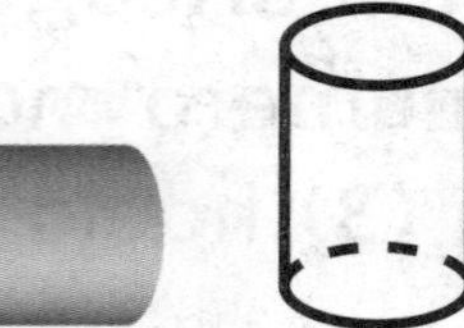
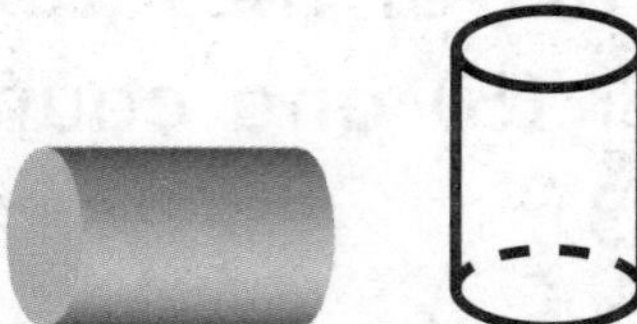

Dd

data Numbers or symbols, sometimes collected from a survey or experiment, that show information. *Data* is plural.

Name	Number of Pets
Mary	3
James	1
Alonzo	4

datos Números o símbolos que se recopilan mediante una encuesta o experimento para mostrar información.

Nombre	Número de mascotas
María	3
James	1
Alonzo	4

day 1 day = 24 hours
Examples: Sunday, Monday, Tuesday, Wednesday, Thursday, Friday, Saturday

día 1 día = 24 horas
Ejemplos: domingo, lunes, martes, miércoles, jueves, viernes y sábado

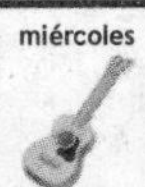

difference The answer to a subtraction problem.

$3 - 1 = 2$

The difference is 2.

diferencia Resultado de un problema de resta.

$3 - 1 = 2$

La diferencia es 2.

digit A symbol used to write numbers. The ten digits are:
0, 1, 2, 3, 4, 5, 6, 7, 8, 9.

dígito Símbolo que se utiliza para escribir números. Los diez dígitos son:
0, 1, 2, 3, 4, 5, 6, 7, 8, 9.

digital clock A clock that uses only numbers to show time.

reloj digital Reloj que marca la hora solo con números.

Dd

dime dime = 10¢ or 10 cents

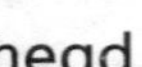

head tail

moneda de 10¢ moneda de diez centavos = 10¢ o 10 centavos

cara cruz

dollar one dollar =100¢ or 100 cents. It can also be written as $1.00 or $1.

front

back

dólar un dólar = 100¢ o 100 centavos. También se puede escribir $1.00.

frente

revés

dollar sign ($) The sign used to show dollars.

one dollar = $1 or $1.00

signo de dólar ($) Símbolo que se usa para mostrar dólares.

un dólar = $1 o $1.00

doubles (and near doubles) Two addends that are the same number.

6 + 6 = 12 ← doubles

6 + 7 = 13 ← near doubles

dobles (y casi dobles) Dos sumandos que son el mismo número.

6 + 6 = 12 ← dobles

6 + 7 = 13 ← casi dobles

Ee

edge The line segment where two *faces* of a three-dimensional shape meet.

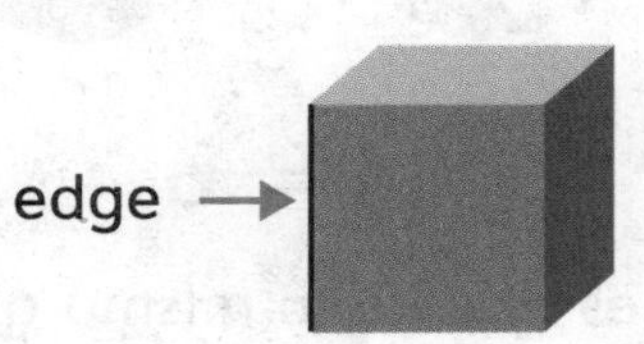

arista Segmento de recta donde se encuentran dos caras de una figura tridimensional.

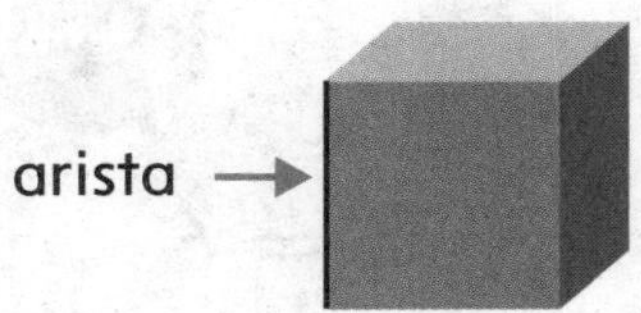

equal groups Each group has the same number of objects.

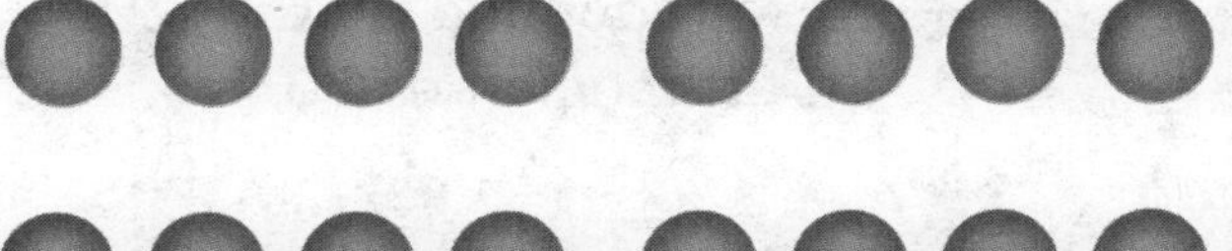

There are four equal groups of counters.

grupos iguales Cada grupo tiene el mismo número de objetos.

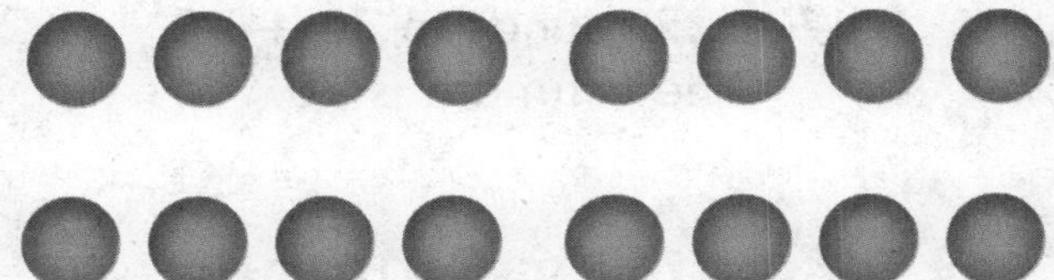

Hay cuatro grupos iguales de fichas.

equal parts Each part is the same size.

This sandwich is cut into 2 equal parts.

partes iguales Cada parte es del mismo tamaño.

Este sándwich está cortado en 2 partes iguales.

Ee

equal to =

6 = 6

6 is *equal to* or the same as 6.

igual a =

6 = 6

6 es *igual* o lo mismo que 6.

estimate Find a number close to an exact amount.

47 + 22 rounds to 50 + 20.
The estimate is 70.

estimar Hallar un número cercano a la cantidad exacta.

47 + 22 se redondea a 50 + 20.
La estimación es 70.

even number Numbers that end with 0, 2, 4, 6, 8.

número par Los números que terminan en 0, 2, 4, 6, 8.

expanded form
The representation of a number as a sum that shows the value of each digit. Sometimes called *expanded notation*.

536 is written as 500 + 30 + 6.

forma desarrollada
La representación de un número como suma que muestra el valor de cada dígito. También se llama *notación desarrollada*.

536 se escribe como 500 + 30 + 6.

Ff

face The flat part of a three-dimensional shape.

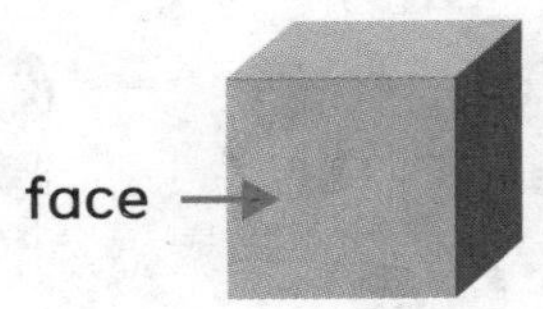

A square is a face of a cube.

cara La parte plana de una figura tridimensional.

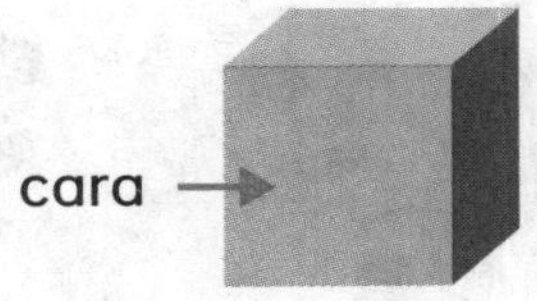

El cuadrado es la cara de un cubo.

fact family Addition and subtraction sentences that use the same numbers.

$6 + 7 = 13$ $13 - 7 = 6$
$7 + 6 = 13$ $13 - 6 = 7$

familia de operaciones Enunciados de suma y resta los cuales tienen los mismos números.

$6 + 7 = 13$ $13 - 7 = 6$
$7 + 6 = 13$ $13 - 6 = 7$

foot (ft) A customary unit for measuring length. Plural is feet.

1 foot = 12 inches

pie Unidad usual para medir longitud.

1 pie = 12 pulgadas

fourths Four equal parts of a whole. Each part is a fourth, or a quarter of the whole.

cuartos Cuatro partes iguales de un todo. Cada parte es un cuarto, o la cuarta parte del todo.

Gg

greater than >

$$7 > 2$$

7 is greater than 2.

mayor que >

$$7 > 2$$

7 es mayor que 2.

group A set of objects.

1 group of 4

grupo Conjunto o grupo de objetos.

1 grupo de 4

half hour (or half past) A unit to measure time. Sometimes called *half past* or *half past the hour.*

a half hour = 30 minutes

media hora (o y media) Unidad para medir tiempo. A veces se dice *hora y media.*

media hora = 30 minutos

halves Two equal parts of a whole. Each part is a half of the whole.

mitades Dos partes iguales de un todo. Cada parte es la mitad de un todo.

hexagon A 2-dimensional shape that has six sides.

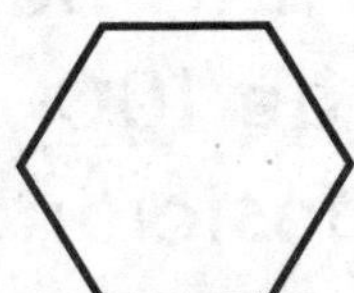

hexágono Figura bidimensional que tiene seis lados.

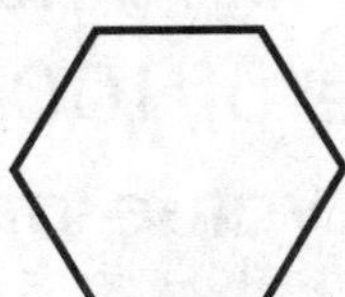

hour A unit to measure time.

I hour = 60 minutes

hora Unidad para medir tiempo.

I hora = 60 minutos

hour hand The hand on a clock that tells the hour. It is the shorter hand.

manecilla horaria Manecilla del reloj que indica la hora. Es la más corta.

hundreds The numbers in the range of 100–999. It is the place value of a number.

365

3 is in the hundreds place.
6 is in the tens place.
5 is in the ones place.

centenas Los números en el rango de 100 a 999. Es el valor posicional de un número.

365

3 está en el lugar de las centenas.
6 está en el lugar de las decenas.
5 está en el lugar de las unidades.

Ii

inch (in) A customary unit for measuring length. The plural is *inches*.

12 inches = 1 foot

pulgada (pulg) Unidad usual para medir longitud.

12 pulgadas = 1 pie

inverse Operations that are opposite of each other.

Addition and subtraction are inverse or opposite operations.

operaciones inversas Operaciones que se oponen una a otra.

La suma y la resta son operaciones inversas u opuestas.

Kk

key Tells what (or how many) each symbol stands for.

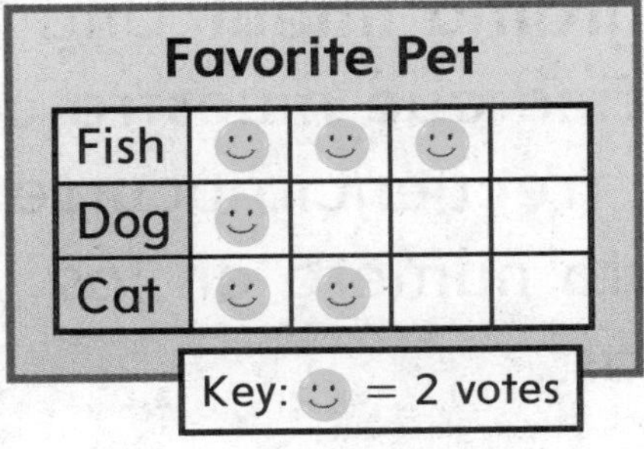

Favorite Pet

Fish	☺	☺	☺	
Dog	☺			
Cat	☺	☺		

Key: ☺ = 2 votes

clave Indica qué o cuánto representa cada símbolo.

Animal doméstico favorito

Pez	☺	☺	☺	
Perro	☺			
Gato	☺	☺		

Clave: ☺ = 2 votos

Ll

length How long or how far away something is.

longitud El largo de algo o lo lejos que está.

less than <

$4 < 7$

4 is less than 7.

menor que <

$4 < 7$

4 es menor que 7.

line plot A graph that shows how often a certain number occurs in data.

diagrama lineal Una gráfica que muestra con qué frecuencia ocurre cierto número en los datos.

Mm

measure To find the length, height, weight, capacity, or temperature using standard or nonstandard units.

medir Hallar la longitud, altura, peso, capacidad o temperatura usando unidades estándares o no estándares.

meter (m) A metric unit for measuring length.

1 meter = 100 centimeters

metro (m) Unidad métrica para medir longitud.

1 metro = 100 centímetros

minute (min) A unit to measure time. Each tick mark is one minute.

1 minute = 60 seconds

minuto (min) Unidad para medir tiempo. Cada marca es un minuto.

1 minuto = 60 segundos

minute hand The longer hand on a clock that tells the minutes.

minutero La manecilla más larga del reloj. Indica los minutos.

Mm

missing addend The missing number in a number sentence that makes the number sentence true.

$9 + \square = 16$

The missing addend is 7.

sumando desconocido El número desconocido en un enunciado numérico que hace que este sea verdadero.

$9 + \square = 16$

El sumando desconocido es 7.

month A unit of time.
12 months = 1 year

April

Sunday	Monday	Tuesday	Wednesday	Thursday	Friday	Saturday
		1	2	3	4	5
6	7	8	9	10	11	12
13	14	15	16	17	18	19
20	21	22	23	24	25	26
27	28	29	30			

This is the month of April.

mes Unidad de tiempo.
12 mesas = 1 año

Abril

domingo	lunes	martes	miércoles	jueves	viernes	sábado
		1	2	3	4	5
6	7	8	9	10	11	12
13	14	15	16	17	18	19
20	21	22	23	24	25	26
27	28	29	30			

Este es el mes de abril.

Nn

near doubles Addition facts in which one addend is exactly 1 more or 1 less than the other addend.

casi dobles Operaciones de suma en las cuales un sumando es exactamente 1 más o 1 menos que el otro sumando.

nickel nickel = 5¢ or 5 cents

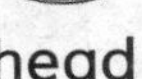
head

tail

moneda de 5¢ moneda de cinco centavos = 5¢ o 5 centavos

cara

cruz

number line A line with number labels.

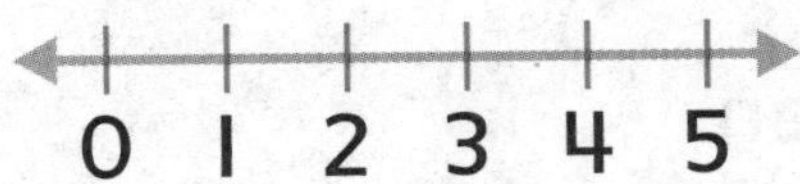

recta numérica Recta con marcas de números.

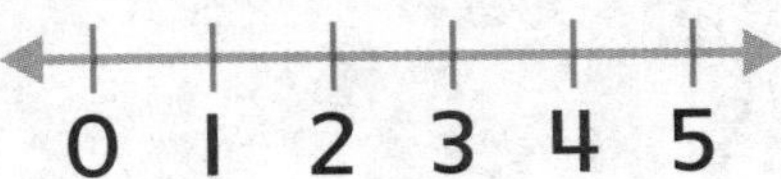

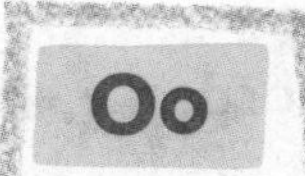

o'clock At the beginning of the hour.

It is 7 o'clock.

en punto El momento en que comienza cada hora.

Son las 7 en punto.

odd number Numbers that end with 1, 3, 5, 7, 9.

números impares Los números que terminan en 1, 3, 5, 7, 9.

Oo

ones The numbers in the range of 0-9. A place value of a number.

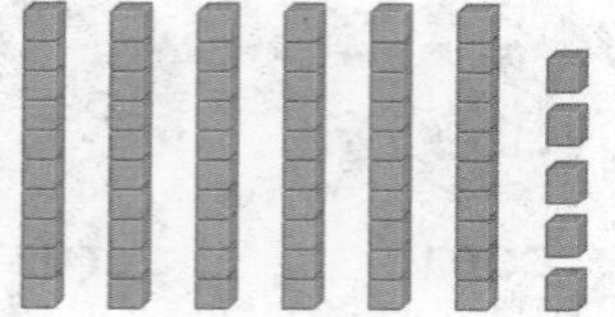

65

5 is in the ones place.

unidades Los números en el rango de 0 a 9. Valor posicional de un número.

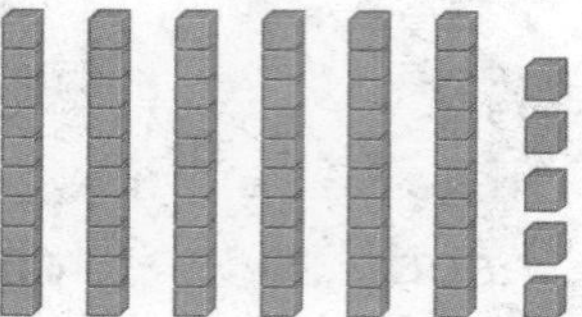

65

El 5 está en el lugar de las unidades.

order

I, 3, 6, 8, 10

These numbers are in order from least to greatest.

orden

I, 3, 6, 8, 10

Estos números están en orden de menor a mayor.

Pp

P.M. The hours from noon until midnight.

p.m. Las horas que van desde el mediodía hasta la medianoche.

parallelogram A two-dimensional shape that has four sides. Each pair of opposite sides is equal and parallel.

paralelogramo Figura bidimensional que tiene cuatro lados. Cada par de lados opuestos son iguales y paralelos.

partition To divide or "break up."

separar Dividir o desunir.

pattern An order that a set of objects or numbers follows over and over.

patrón Orden que sigue continuamente un conjunto de objetos o números insertar punto.

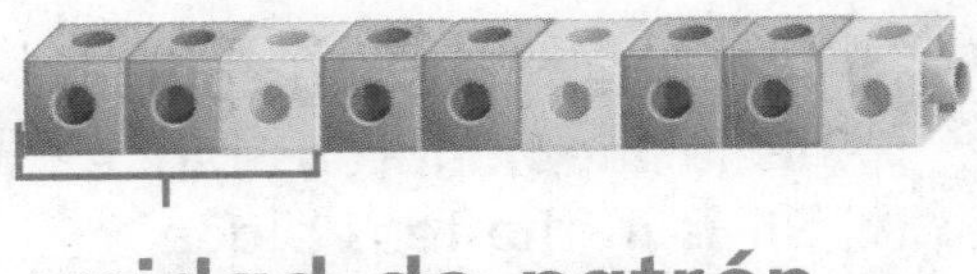

penny penny = 1¢ or 1 cent

head

tail

moneda de 1¢ moneda de un centavo = 1¢ o 1 centavo

cara

escudo

pentagon A polygon with five sides.

pentágono Polígono de cinco lados.

picture graph A graph that has different pictures to show information collected.

How I Get to School

Bus									
Bike									
Walk									

gráfica con imágenes Gráfica que tiene diferentes imágenes para ilustrar la información recopilada.

Cómo voy a la escuela

En Autobús									
En Bicicleta									
Caminando									

Pp

place value The value given to a *digit* by its place in a number.

1,365

1 is in the thousands place.
3 is in the hundreds place.
6 is in the tens place.
5 is in the ones place.

valor posicional El valor dado a un *dígito* según su posición en un número.

1,365

1 está en el lugar de los millares.
3 está en el lugar de las centenas.
6 está en el lugar de las decenas.
5 está en el lugar de las unidades.

pyramid A three-dimensional shape with a polygon as a base and other faces that are triangles.

pirámide Figura tridimensional con un polígono como base y otras caras que son triángulos.

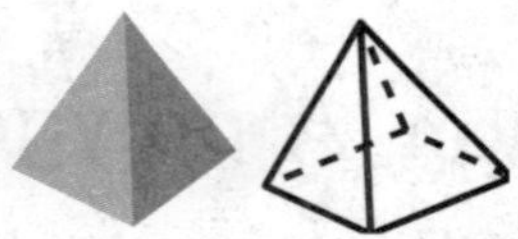

Qq

quadrilateral A shape that has 4 sides and 4 angles.

cuadrilátero Figura con 4 lados y 4 ángulos.

quarter quarter = 25¢ or 25 cents

head

tail

moneda de 25¢ moneda de 25 centavos = 25¢ o 25 centavos

cara

cruz

quarter hour A quarter hour is 15 minutes. Sometimes called *quarter past* or *quarter til.*

cuarto de hora Un cuarto de hora es 15 minutos. A veces se dice hora y cuarto.

rectangle A plane shape with four sides and four corners.

rectángulo Figura plana con cuatro lados y cuatro esquinas.

rectangular prism A three-dimensional shape with 6 faces that are rectangles.

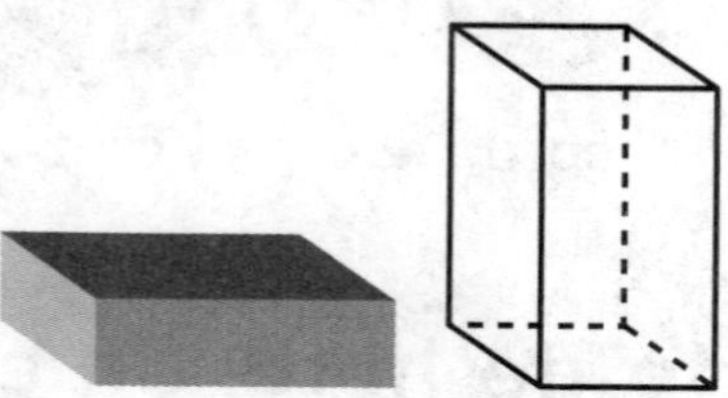

prisma rectangular Figura tridimensional con 6 caras que son rectángulos.

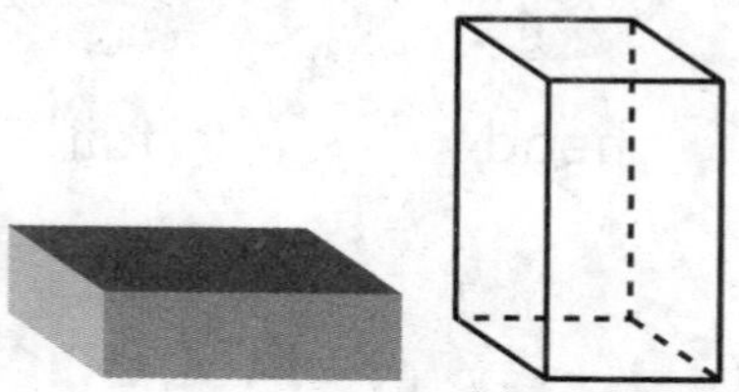

regroup Take apart a number to write it in a new way.

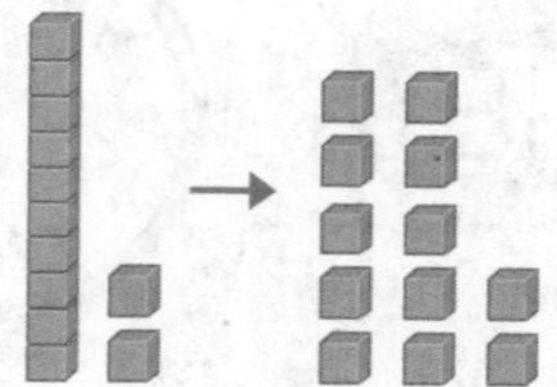

1 ten + 2 ones becomes 12 ones.

reagrupar Separar un número para escribirlo de una nueva forma.

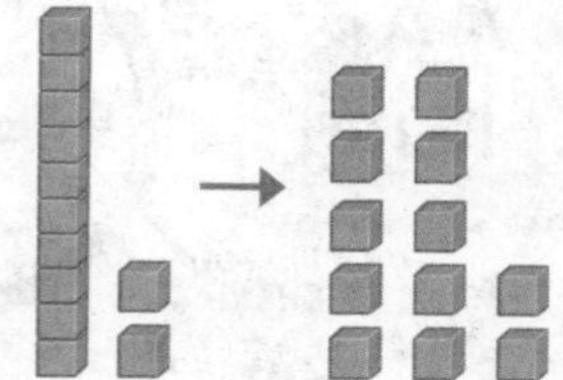

1 decena + 2 unidades se convierten en 12 unidades.

related fact(s) Basic facts using the same numbers. Sometimes called a *fact family*.

$4 + 1 = 5$ $5 - 4 = 1$

$1 + 4 = 5$ $5 - 1 = 4$

operaciones relacionadas Operaciones básicas en las que se usan los mismos números. También se llaman *familias de operaciones*.

$4 + 1 = 5$ $5 - 4 = 1$

$1 + 4 = 5$ $5 - 1 = 4$

repeated addition To use the same addend over and over.

suma repetida Usar el mismo sumando una y otra vez.

rhombus A shape with 4 sides of the same length.

rombo Paralelogramo con cuatro lados de la misma longitud.

round Change the *value* of a number to one that is easier to work with.

24 rounded to the nearest ten is 20.

redondear Cambiar el *valor* de un número a uno con el que es más fácil trabajar.

24 redondeado a la decena más cercana es 20.

Ss

side One of the line segments that make up a shape.

A pentagon has five sides.

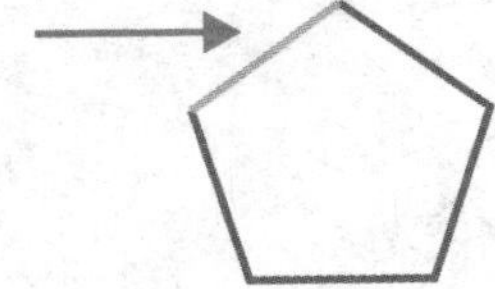

lado Uno de los segmentos de recta que componen una figura.

El pentágono tiene cinco lados.

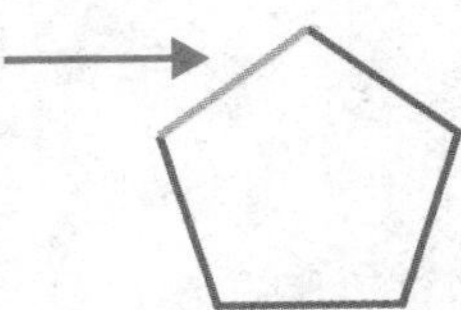

skip count Count objects in equal groups of two or more.

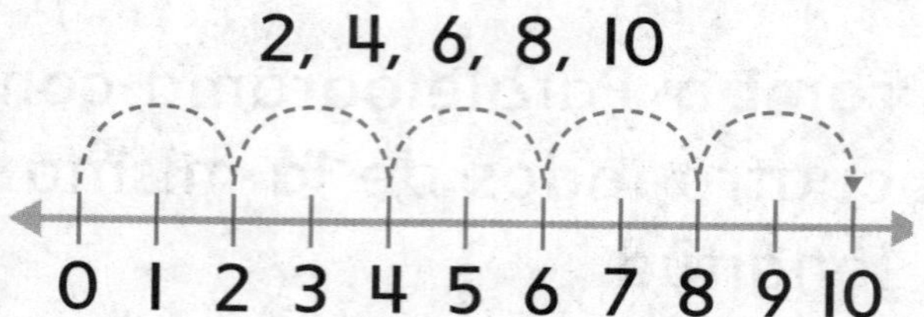

contar salteado Contar objetos en grupos iguales de dos o más.

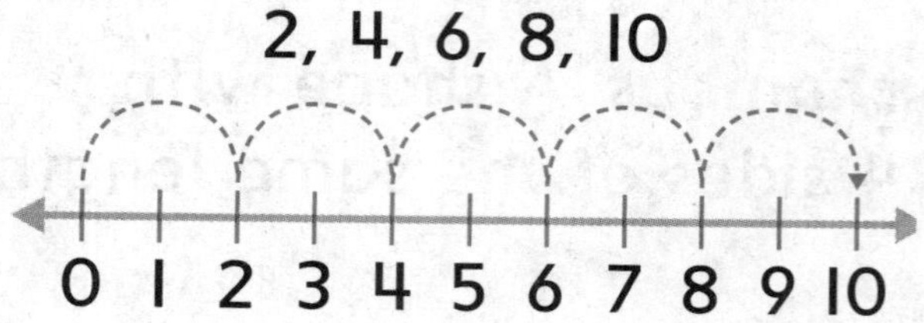

slide To move a shape in any direction to another place.

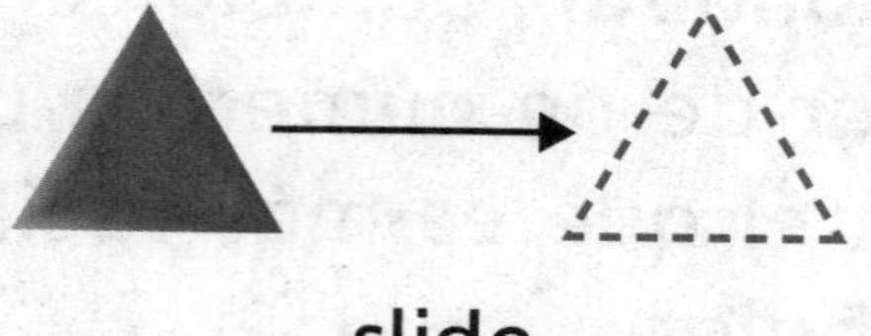

slide

deslizar Traslador una figura a una nueva posición.

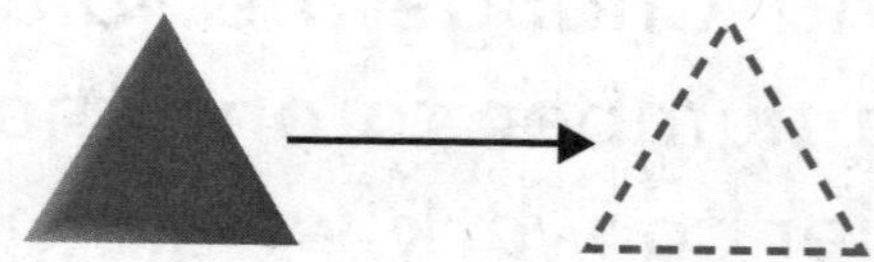

deslizar

sphere A three-dimensional shape that has the shape of a round ball.

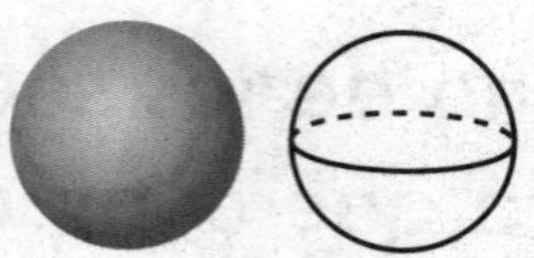

esfera Figura tridimensional que tiene la forma de una pelota redonda.

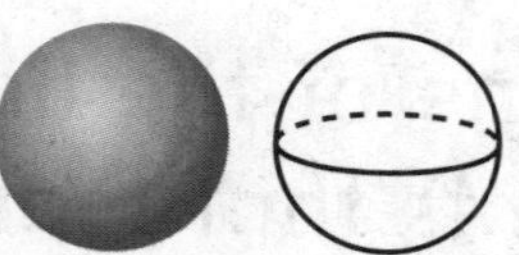

square A two-dimensional shape that has four equal sides. Also a rectangle.

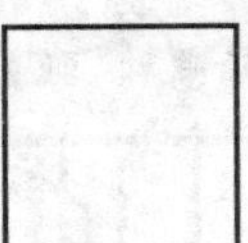

cuadrado Figura bidimensional que tiene cuatro lados iguales. También es un rectángulo.

subtract (subtraction) Take away, take apart, separate, or find the difference between two sets. The opposite of *add*.

$5 - 5 = 0$

restar (sustracción) Eliminar, quitar, separar o hallar la diferencia entre dos conjuntos. Lo opuesto de *sumar*.

$5 - 5 = 0$

sum The answer to an addition problem.

$2 + 4 = 6$

suma Resultado de la operación de sumar.

$2 + 4 = 6$

survey Collect data by asking people the same question.

Favorite Animal	
Dog	𝍸 I
Cat	𝍸

This survey shows favorite animals.

encuesta Recopilar datos al hacer las mismas preguntas a un grupo de personas.

Animal favorito	
Perro	𝍸 I
Gato	𝍸

Esta encuesta muestra los animales favoritos.

symbol A letter or figure that stands for something.

This symbol means to add.

$+$

símbolo Letra o figura que representa algo.

Este símbolo significa sumar.

$+$

Tt

tally marks A mark used to record data collected in a survey.

𝍸 II

tally marks

marca de conteo Marca que se utiliza para registrar los datos recopilados en una encuesta.

𝍸 II

marcas de conteo

tens The numbers in the range of 10–99. It is the place value of a number.

65

6 is in the tens place.
5 is in the ones place.

decenas Los números en el rango de 10 a 99. Es el valor posicional de un número.

65

6 está en el lugar de las decenas.
5 está en el lugar de las unidades.

thirds Three equal parts.

tercios Tres partes iguales.

thousand(s) The numbers in the range of 1,000–9,999. It is the place value of a number.

1,365

1 is in the thousands place.
3 is in the hundreds place.
6 is in the tens place.
5 is in the ones place.

millar(es) Los números en el rango de 1,000 a 9,999. Es el valor posicional de un número.

1,365

1 está en el lugar de los millares.
3 está en el lugar de las centenas.
6 está en el lugar de las decenas.
5 está en el lugar de las unidades.

three-dimensional shape A shape that has length, width, and height.

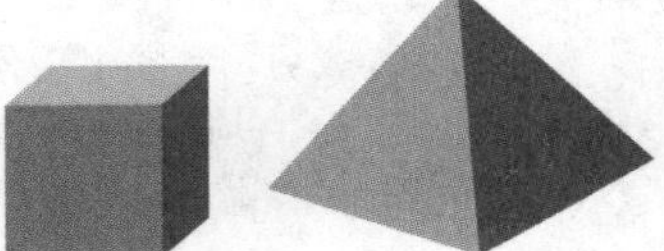

figura tridimensional Que tiene tres dimensiones: largo, ancho y alto.

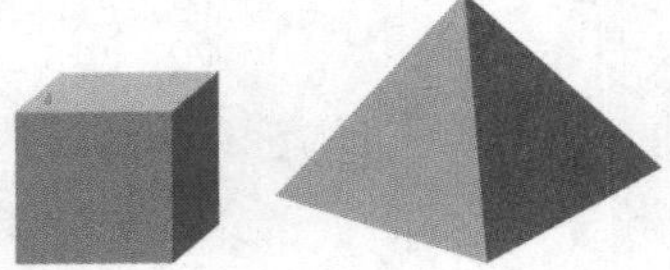

trapezoid A two-dimensional shape with four sides and only two opposite sides that are parallel.

trapecio Figura bidimensional de cuatro lados con solo dos lados opuestos que son paralelos.

triangle A two-dimensional shape with three sides and three angles.

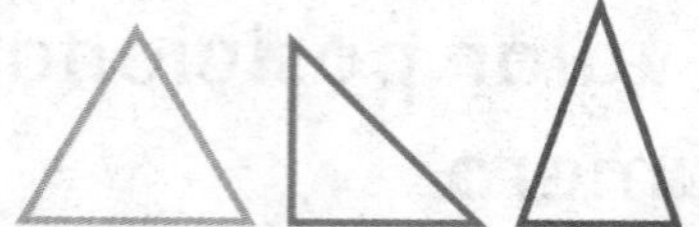

triángulo Figura bidimensional con tres lados y tres ángulos.

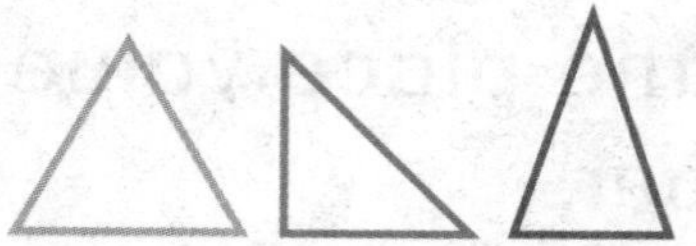

two-dimensional shape The outline of a shape—such as a triangle, square, or rectangle—that has only *length* and *width*.

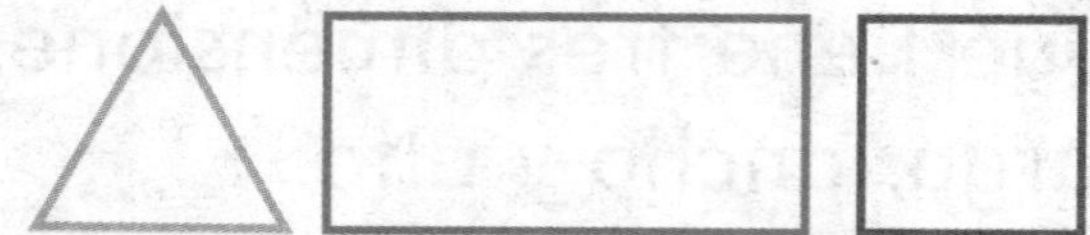

figura bidimensional Contorno de una figura, como un triángulo, un cuadrado o un rectángulo, que solo tiene *largo* y *ancho*.

Vv

vertex

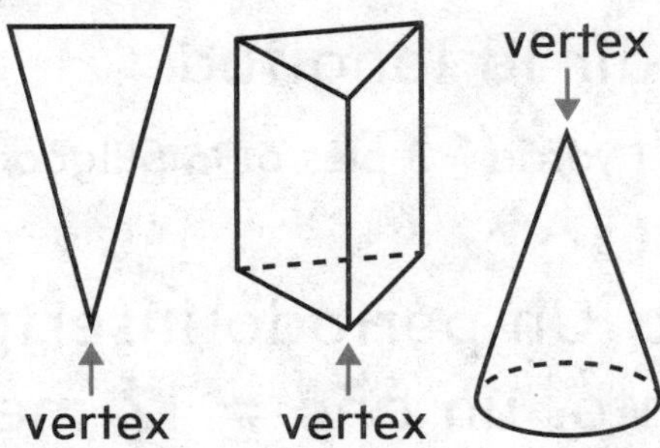

vértice

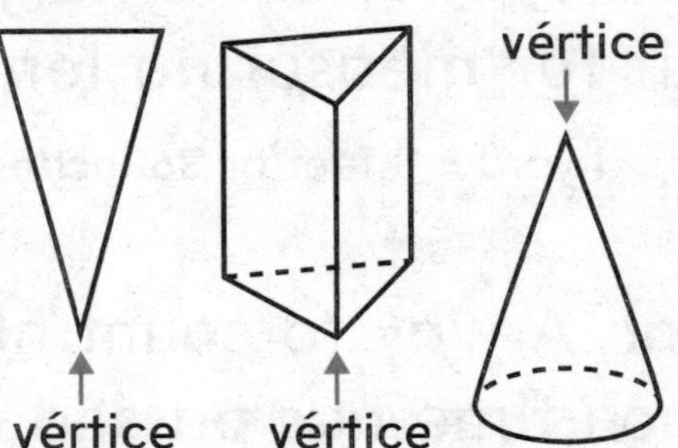

week A part of a calendar.
1 week = 7 days

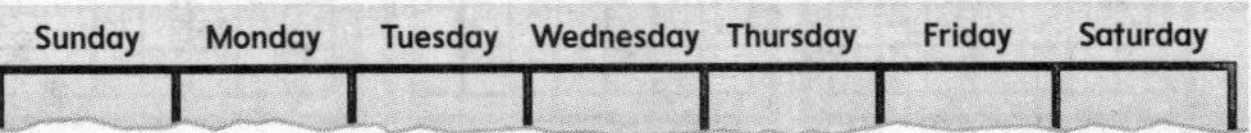

semana Parte de un calendario una semana = 7 días

domingo lunes martes miércoles jueves viernes sábado

whole The entire amount or object.

el todo La cantidad total o el objeto completo.

yard (yd) A customary unit for measuring length.

1 yard = 3 feet or 36 inches

yarda Unidad usual para medir la longitud.

1 yarda = 3 pies o 36 pulgadas

year A way to count how much time has passed or will pass. 1 year = 12 months

January

S	M	T	W	T	F	S
						1
2	3	4	5	6	7	8
9	10	11	12	13	14	15
16	17	18	19	20	21	22
23	24	25	26	27	28	29
30	31					

February

S	M	T	W	T	F	S
		1	2	3	4	5
6	7	8	9	10	11	12
13	14	15	16	17	18	19
20	21	22	23	24	25	26
27	28					

March

S	M	T	W	T	F	S
		1	2	3	4	5
6	7	8	9	10	11	12
13	14	15	16	17	18	19
20	21	22	23	24	25	26
27	28	29	30	31		

April

S	M	T	W	T	F	S
					1	2
3	4	5	6	7	8	9
10	11	12	13	14	15	16
17	18	19	20	21	22	23
24	25	26	27	28	29	30

May

S	M	T	W	T	F	S
1	2	3	4	5	6	7
8	9	10	11	12	13	14
15	16	17	18	19	20	21
22	23	24	25	26	27	28
29	30	31				

June

S	M	T	W	T	F	S
			1	2	3	4
5	6	7	8	9	10	11
12	13	14	15	16	17	18
19	20	21	22	23	24	25
26	27	28	29	30		

July

S	M	T	W	T	F	S
					1	2
3	4	5	6	7	8	9
10	11	12	13	14	15	16
17	18	19	20	21	22	23
24	25	26	27	28	29	30
31						

August

S	M	T	W	T	F	S
	1	2	3	4	5	6
7	8	9	10	11	12	13
14	15	16	17	18	19	20
21	22	23	24	25	26	27
28	29	30	31			

September

S	M	T	W	T	F	S
				1	2	3
4	5	6	7	8	9	10
11	12	13	14	15	16	17
18	19	20	21	22	23	24
25	26	27	28	29	30	

October

S	M	T	W	T	F	S
						1
2	3	4	5	6	7	8
9	10	11	12	13	14	15
16	17	18	19	20	21	22
23	24	25	26	27	28	29
30	31					

November

S	M	T	W	T	F	S
		1	2	3	4	5
6	7	8	9	10	11	12
13	14	15	16	17	18	19
20	21	22	23	24	25	26
27	28	29	30			

December

S	M	T	W	T	F	S
				1	2	3
4	5	6	7	8	9	10
11	12	13	14	15	16	17
18	19	20	21	22	23	24
25	26	27	28	29	30	31

año Un período insertar punto, un año = 12 meses

enero

D	L	M	M	J	V	S
						1
2	3	4	5	6	7	8
9	10	11	12	13	14	15
16	17	18	19	20	21	22
23	24	25	26	27	28	29
30	31					

febrero

D	L	M	M	J	V	S
		1	2	3	4	5
6	7	8	9	10	11	12
13	14	15	16	17	18	19
20	21	22	23	24	25	26
27	28					

marzo

D	L	M	M	J	V	S
		1	2	3	4	5
6	7	8	9	10	11	12
13	14	15	16	17	18	19
20	21	22	23	24	25	26
27	28	29	30	31		

abril

D	L	M	M	J	V	S
					1	2
3	4	5	6	7	8	9
10	11	12	13	14	15	16
17	18	19	20	21	22	23
24	25	26	27	28	29	30

mayo

D	L	M	M	J	V	S
1	2	3	4	5	6	7
8	9	10	11	12	13	14
15	16	17	18	19	20	21
22	23	24	25	26	27	28
29	30	31				

junio

D	L	M	M	J	V	S
			1	2	3	4
5	6	7	8	9	10	11
12	13	14	15	16	17	18
19	20	21	22	23	24	25
26	27	28	29	30		

julio

D	L	M	M	J	V	S
					1	2
3	4	5	6	7	8	9
10	11	12	13	14	15	16
17	18	19	20	21	22	23
24	25	26	27	28	29	30
31						

agosto

D	L	M	M	J	V	S
	1	2	3	4	5	6
7	8	9	10	11	12	13
14	15	16	17	18	19	20
21	22	23	24	25	26	27
28	29	30	31			

septiembre

D	L	M	M	J	V	S
				1	2	3
4	5	6	7	8	9	10
11	12	13	14	15	16	17
18	19	20	21	22	23	24
25	26	27	28	29	30	

octubre

D	L	M	M	J	V	S
						1
2	3	4	5	6	7	8
9	10	11	12	13	14	15
16	17	18	19	20	21	22
23	24	25	26	27	28	29
30	31					

noviembre

D	L	M	M	J	V	S
		1	2	3	4	5
6	7	8	9	10	11	12
13	14	15	16	17	18	19
20	21	22	23	24	25	26
27	28	29	30			

diciembre

D	L	M	M	J	V	S
				1	2	3
4	5	6	7	8	9	10
11	12	13	14	15	16	17
18	19	20	21	22	23	24
25	26	27	28	29	30	31

Name ....

Work Mat 3: Number Lines

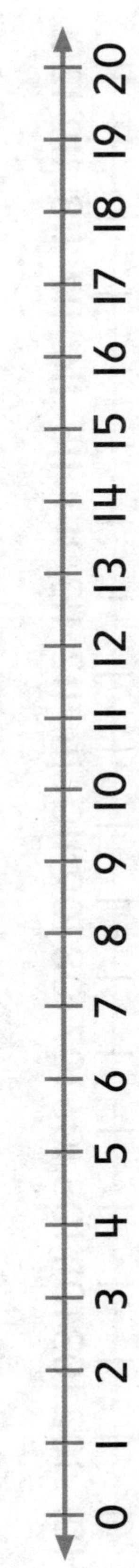

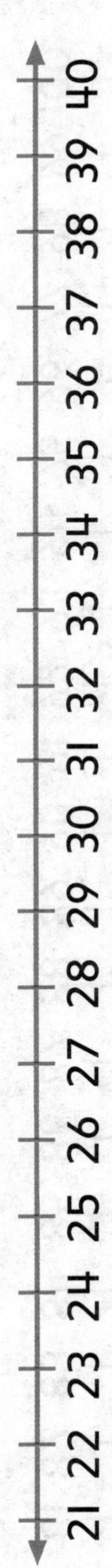

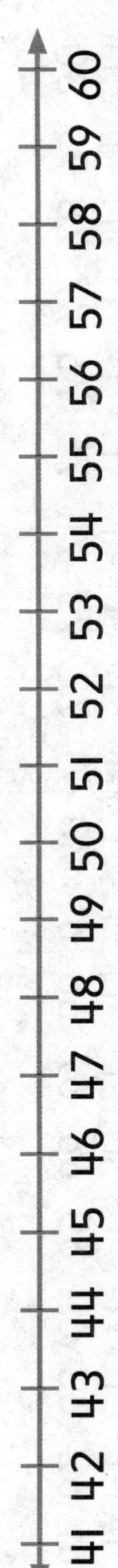

Work Mat 4: Number Lines

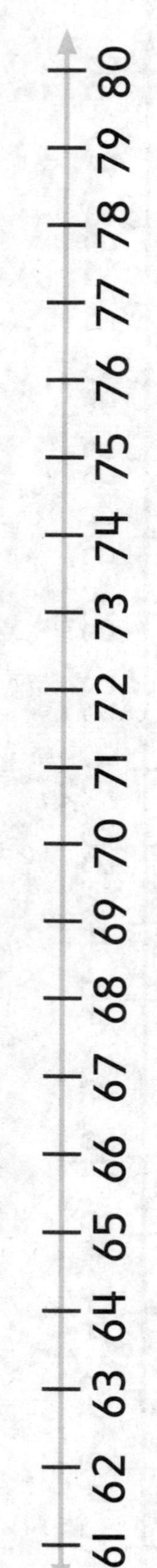

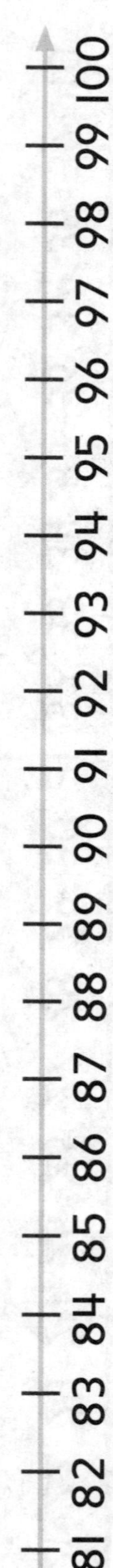

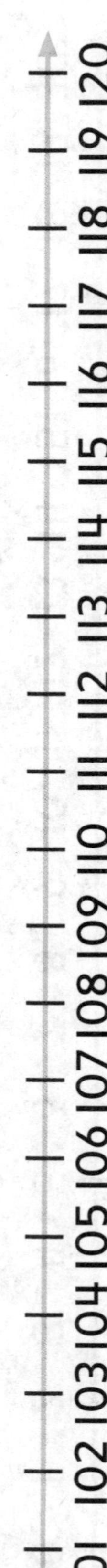